CONFLUENCE

A Tom Armstrong

Investigative Series

By

Stephen Rodgers

Publisher of Imprint: Sage Center Publications

Cover Design by Tana Sollars

All characters in this book are fictional unless I have their permission. These would include my son, his girlfriend (created the cover), and his friend and business partner. Clint, one of the protagonists, is the son of my oldest friend who has appeared in other books of mine. His wife, Luly, is, in real life, Clint's fiancée.

The Rev. Stephen Rodgers is an Episcopal priest and a psychotherapist. He has written books on a wide range of subjects that you can see below. He lives in a floating home on a tributary of the Columbia River.

Voice of the Soul - A book on what the soul is, how it's different from the mind, and how to help it grow. This is a unique theory about how the soul works and how to find it.

Addiction and Recovery: A Practical Guide - written with Dr. Brian Esparza - An easy to read book that explains what goes on in the brain, neurobiologically, how addiction starts, and how to stay in recovery for life.

Alchemical Assistance - A Christian murder thriller that takes place in Portland, Oregon.

The Church Calendar: A Guide to the Seasons of the Church Year - This book is filled with information about the history, liturgy, scripture, spirituality, traditions, theology, and activities around the seasons of the Church calendar.

The Phoenix Too Shall Rise - A 150-year-old alchemist teaches a sixteen-year-old girl about the power of alchemy. In the process, they get into a lot of mischief and discover things about themselves and the universe around them.

Lake River Almanac - Bringing together nature, science, psychology, and spirituality. This book weaves all of these together as they are often a reflection of each other.

God Arrives: The Choice Is Yours – This is an alternative view of the second coming of Christ that differs greatly from the typical – and is based on biblical principles.

Dedication

I wish to dedicate this book to the love of my life, Carol. She inspires and challenges me and fills me with joy. I enjoy being on the journey of life with her. She was also very supportive on many levels with the writing of this book.

Acknowledgments

I want to thank Amy Koski, the Center Manager and Clinical Research Coordinator for the Center for Embryonic Cell and Gene Therapy at the Oregon Health and Science University, for looking over the genetic components of the book.

Bart Massy, Ph.D., a professor at Portland State University, helped me with the AI portions of the book. His assistance helped correct errors in my technical knowledge. He graciously gave comments on other parts of the manuscript that were very helpful.

If you find factual errors, they are not Amy or Bart's fault. I have certainly taken some license with both AI and GE.

While this book may come across as pessimistic, I am an optimist. I believe in our system of governance and that most politicians are decent people. I am a believer in our military and that there are far more amazing people than bad apples. Technology is a remarkable tool, and we all know it can be used for good and evil. On all these levels, we have to keep watch.

Prologue

We live in frightening times. The news is filled with dictators making threats, an economy that rises and falls at times with news that isn't even real. The United States, and other countries, have dealt with mass shootings, pandemics, protests, and a political system where listening, compromising, common sense, and civility have gone out the window – on both sides of the aisle.

If we are honest, all times have something to be a bit fearful of. Wars, disease, oppression, economic collapse, and severe weather are but a few of the issues we contend with. Overall, humanity has done well. We move forward. Progress is difficult to slow, and perhaps it should be.

Our time in history is unique – to this point. Think about the things that have changed history over the past 30,000 years. Most of the big ones, like the domestication of animals, agriculture, and transportation, each took hundreds, if not thousands, of years to accomplish on a global scale. In the past 140 years, every one of those has changed dramatically. During the past forty years, that pace has exploded. That doesn't even include flight, communication, medicine, and media.

The advent of transistors and semiconductors has shifted the future of the planet. We are no longer walking into the future; we are galloping.

Some will contend that those of us who have pragmatic, ethical, and moral concerns are naïve or not looking at the facts. I will agree that. Depending on who you ask, you can find "facts" to prove just about anything you want. "Science" is, in reality, often not what it claims to be. In recent history, we had the Covid-19 incident. We heard it was OK not to wear masks, then we had to wear masks. We heard forecasts of vastly different numbers of potential infections and deaths. All in the name of science. When science can't get its act together, people start to fall into the "don't cry wolf" belief.

None of us know the future until it arrives. What if the people who believe climate change is not real are wrong? Even they will admit there is a tipping point of no return. What about the Covid-19 pandemic. Had no one worn masks, what might have been the outcome? Would it have been worth the risk? Did we want to risk killing a billion people on a guess? The pandemic period of 1918 that killed up to 500 million people did not have the technology or the communication systems we have. Fortunately, most of the world was united to stop the spread of Covid-19. The spread occurred primarily because people did stupid things, like not wearing masks. That self-centeredness is at the root of most of the world's issues. It always has been.

Into that world comes Genetic Engineering and Artificial Intelligence. This book is about what happens when those fields reach the next level and overlap. As the reader knows, this took place in the not too distant past. I write about it to fill in the gaps of what has not been made public. Were people in this book not given full immunity, you would not be reading this.

I have written most of this in the first person and using the present tense. It all happened in the near past, and I could not have been present at everything that takes place in this book. I fill in the gaps by talking with all the players and doing a great deal of research, for which I am best known.

I do not trust our government very much, and others far less. What I write about, I have no doubt will happen. It's not a matter of if, but when.

1

Over the past fifteen years, I'd helped put two Senators, a Congressman, two CEO's, and four hedge fund managers behind bars for a massive money laundering and an insider trading scheme. They were not happy. During this investigation, along with two others I'd done, my house had been burned to the ground, my life threatened numerous times, and attempts to tarnish my reputation were almost constant. I was also put in jail twice for not compromising a source.

My name is Tom Armstrong, and I write books about investigations I do about the rich and powerful. I started by writing novels, mostly murder thrillers, but become more involved with investigating reality. I also look into improprieties of our government. There is a great deal to investigate on both fronts. Being in this line of work with the success I have found comes with the good, the bad, and the ugly. The good aspect of building the reputation I have is that I get "leaks" all the time. Fortunately, there are many more good people in the United States, and around the world, than bad. I am adamant about not revealing sources. That is a staple in this trade.

The way I write books is to be penning them as the events unfold. As dots are connected along the way, I share them. The end is not known until it arrives.

I was researching two projects. Many of my friends are teachers, and the evidence clearly shows American education is not doing well. I wanted to dig deeper to find out when it started to slide and why. My work always ended with some solutions. Contacting the best minds in the world, along with political leaders, had led to some significant shifts on several fronts. I was also working on election fraud, international interference, and how states kept people from voting. The corruption on all sides appears to be massive.

I heard the phone ring in the other room. Most of me didn't want to answer, but my brain tweaked, and I got off my ass and picked it up.

"Hello, this is Tom."

"Is this Mr. Armstrong?" said a nervous female voice.

"Yes, it is. Who is this, please?"

People often called me with the latest "scoop." I'd had enough success that I hired someone to run interference. They would listen to the caller's story, and if they deemed it worthy, they put the call through. Less than five percent got through to me.

"I'd rather not say at this point if you don't mind."

"Fair enough, what can I do for you?"

"How much research have you done on biological warfare?" said the woman.

"Barely enough to be dangerous," I replied.

"In 1972, President Nixon unilaterally signed the Biological and Toxin Weapons Convention, which outlawed the United States creating chemical and biological weapons. Because the laboratories are run either by the military or are under the auspices of the National Intelligence Department, there is no independent oversight, none. Let's just say that convention is bending, perhaps has broken."

"You mean we are creating biological weapons?"

"I didn't say that. I am a bit paranoid and don't trust that your phone is not being tapped."

"You are wise, so how can we meet?"

"When can you meet?"

"In two days?"

"That works, go to the Courtyard Marriot on Broadway at 54th. Make sure you get into the elevator alone – other than me. I will be there at 11:45 am."

"How will I know you?"

"Trust me. Try not to be followed."

"They won't have to follow me if they are listening. They know where we are going."

"True, just be in the elevator alone. If no one gets in with you, think about your work, and you'll know what floor."

I felt goosebumps. I was aware that many countries had broken these treaties. I did some fast homework and found the Geneva Protocol was signed in 1925 by most major countries other than the United States. We finally signed in 1974. This banned the use of bacteriological weapons. One hundred countries had signed the Biological Weapons Convention in 1972, including the U.S. It was a known fact that Russia, China, Syria, Iraq, and Iran had created these types of weapons. Most people assumed we were doing something, but there was no proof. In 1989 the U.S. created the Biological Weapons Anti-Terrorism Act of 1989. That law defined a biological agent as:

"any micro-organism, virus, infectious substance, or biological product that may be engineered as a result of biotechnology, or any naturally occurring or bioengineered component of any such microorganism, virus, infectious substance, or biological product, capable of causing death, disease, or other biological malfunction in a human, an animal, a plant, or another living organism; deterioration of food, water, equipment, supplies, or material of any kind ..."

This legislation was amended in 1996 to close some loopholes and again as part of the Patriot Act of 2001.

Calling contacts to get information was a key part of what I did. I had numerous contacts in virtually every department of the government. For some, I'm sure it was the adage that one should keep enemies close. I wanted to get into this but knew I was jumping the gun. I could feel my adrenaline pumping. My instincts were usually right, and I had a bad feeling about this.

2

The following day, I put my other project in a drawer and focused on the new project. I made a list of questions to ask and people to call after the meeting. In all my research, I always had a mental, ethical discussion with myself and with a few of my friends once I was into the work. I wanted to do my best to ensure I wasn't doing more harm than good.

What if the U.S. is making biological weapons or even just researching them? If everyone else is, shouldn't we be? If there were a war and the enemy had biological weapons, and we didn't, wouldn't that be a bad thing?

Nixon had defunded biological research for offensive weapons. He felt they were overrated. He continued to support those laboratories for defensive purposes. They could find ways of protecting us against biological and chemical weapons but not create them. The core bio-agents they worked with gave birth to the diseases of anthrax, tularemia, brusellosis, Q-fever, VEE, botulism, and of course, smallpox. Over time the list grew to include AIDS, Ebola, SARS, and Covid-19.

In the United States, close to $150 billion is part of the "black budget." This means there is no notification about where or how the money is spent. Research, covert military operations, the CIA, and the National Intelligence Agency all get a piece. Obama increased the black budget by 30% in 2009. Trump added another 30%. Since then, another 45% has been added.

This research can't be done just anywhere. Because of the potential for disaster, all bioagents mentioned earlier (and many others) must be kept in a BSL-4 (Biosafety Level) laboratory. There are 15 in the U.S. and who knows how many around the world. All of those in the U.S. are owned by the government, except for a private lab in Texas. Several are with universities.

While conspiracy theorists believe Covid-19 was created by the Chinese, the overwhelming evidence is that it was a natural mutation, probably in bats. What if a nation did create a virus? Would it not be ethical to make the information public that a country had such a virus? Did anyone doubt for a second that many countries were trying to find such a beast?

It is difficult to believe that, if hundreds or even just dozens of scientists were involved in creating such a virus, they could all keep it secret. This is true of conspiracy theories about UFOs in Area 51, landing on the moon, and the killing of JFK. On the other hand, if you and your family were threatened, you might.

With my other projects, I did not always reveal everything I'd found. You have to be careful making public highly classified materials. Vindication, arrests, and death are not uncommon in the world, including the government.

Let's be clear. I value my life. Investigating that our government was potentially creating human-made viruses scared me. So did the fact that I doubted the government would knowingly let me publish that information, no matter the cost. I could feel the sweat already forming on my brow. Yes, I am prone to exaggeration and paranoia at this stage of a project. Normally, I look at myself in the mirror and say, "Tom, you are an idiot, get over yourself." I was having a more difficult time doing that this time.

I spent an hour trying to figure out what was needed to actually create a virus. When the U.S. intelligence looked at Iran and its nuclear potential, we knew what Iran would need to make that happen. We could make predictions about how long it would take once they had those tools and resources. The issue with viruses was that we knew what was needed; we just didn't know what countries had those tools and resources. If history was any indication, we know some countries will not find it necessary to do this type of work in a level 4 laboratory. They don't really care about their citizens. In fact, in Russia, China, and Germany, citizens were used as guinea pigs until they found others, like U.S. POWs and those they were going to kill anyway. In the 1950s, prisoners and soldiers were used as guinea pigs for the military and CIA.

The internet had changed the world. Virtually any country could build a lab and acquire the resources to develop a virus. If that scares you, it should. It is not a question of if, but when. The questions then become what will they do with it, who will be the guinea pigs, will they create a vaccine or treatment at the same time, and what is the end game?

If your country was threatened with almost complete annihilation, would you turn over power? Would you give half your GDP as a ransom to acquire the vaccine? Do first world countries care if this were to happen to a third world country? Would this lead to another "bio-race" like the arms race? Would we enter another cold war based on biological MAD - mutually assured destruction? The questions were endless.

I had a long list of journalists who'd looked into biological warfare. I added scientists, professors, military personnel, and people in the political arena that I knew. By the time I was done, I had over 85 names.

Being a known quantity had its drawbacks. Once word hit the street that I was investigating biological warfare, doors would close, calls would not be returned, and the work would become more difficult, if not treacherous. I started to organize the list with those who were most likely to spread the word at the bottom. Having a potential mole was always a help. Watergate had its "Deep Throat." Eventually, I'd need to come up with a nickname for mine. Now I was jumping ahead of the game. Slow down.

The next day, I left my home and headed out. I have been followed on a few of my investigations. I think it was more a method to instill fear into my life, so they made it obvious. I didn't think anyone would be watching today, but I wanted to honor the informant's paranoia.

The Courtyard Marriot in New York is one of the tallest hotels in the city. It has sixty-eight floors and 378 rooms and 261 suites. The lobby is spacious and spectacular. The hotel is located in mid-town, not far from Times Square. If someone was following me, they were already here. I looked around to see if they were making it apparent that I was being followed. No one walked up and told me.

There were numerous elevators. I thought an Asian tourist group had just arrived since a couple of dozen people were waiting to get on. I was bumping into people, hoping for an empty elevator. A woman bumped into me and, as she moved on, stuffed a piece of paper in my hand. My guess was she was the contact. I walked around and took out the note.

It said: "Take an elevator to the 34th floor. Get out, get back in an empty one, and go to the 52nd floor."

This all seemed a bit much, but I did as directed. Fortunately, the 34th floor was quiet, and it was easy to find an elevator that was empty. I got out, and there was the same young woman standing in front of me.

"Hi, I'm Tom."

"I know, follow me."

We walked across and got into the next elevator that arrived. There was one woman in it. She got off at the lobby, and we continued to the parking lot two stories down. A car was waiting.

"I called Uber," she said. We got in, and she gave directions to a location about a mile away.

"I apologize for my paranoia, but I think you understand."

"It seems a bit much at this stage of things, but I respect your feelings."

3

"Thank you."

"Can I know your name?"

"Mary."

"Mary...what?"

"Can we leave it at Mary for now?"

"Of course."

There was more silence than talking. Mary clearly wanted it that way, and I didn't push the conversation.

"Here we are," she said, getting out. It was an old Co-op complex. The doorman opened the door and welcomed her. We went to the thirteenth floor.

"Is this your place?" I asked.

"No, my grandmother's, she is away for a week."

We entered her apartment. It didn't take long to see this belonged to someone in their 80's or 90's. Old paintings and pictures, lots of family photos, and numerous pieces of antique furniture.

"Can I get you something?"

"A glass of water would be fine, thank you."

I sat in the living room, taking it all in. She returned with my water and her tea.

"Thank you for your willingness to talk with me and do what I asked to have this meeting."

"Do not worry about that. Tell me what is on your mind that is causing this nervousness."

I finally stopped to look at her. I'd been too preoccupied with everything else to notice. She was probably thirty-five, although I am not the best at gauging the ages of people. It was summer, and she was dressed casually in a summer dress, sandals, and a baseball hat – the Washington Nationals. My guess was she biked a lot. Her body was fit, her posture almost perfect, and I saw a picture on the mantel of her with a racing bike. I saw no wedding ring or any jewelry. She was quite attractive. In my line of work, you didn't get into personal discussions for a long time, if ever.

"So, I am assuming I can trust you. My name will never be used without my permission, and you will keep it confidential."

"Yes. I have gone to jail a few times, keeping someone's name confidential. I don't even remember who they are."

We both laughed. It was always wise to try to put the source's mind at ease. Humor was an excellent way to start.

"I will try to be brief. Stop me anytime you have a question. If I say something you already know about, be blunt and tell me to move on. I don't want to waste your time."

"Understood. Can I take notes or record you?"

"Either is fine."

I pulled out my laptop and started to record.

"I would prefer the first part not to be recorded or written down. I am going to tell you who I am, where I work, and what I do. I would also hope you would not write that information down after we have met. That is my paranoia."

"It's good to be careful. I create codenames for everyone I talk with who is in your type of situation. Your name and other personal information will never be written down."

"Thanks. I guess you've done this a few times."

"A few."

"I called you because of who you are and the work you have done. You are an excellent investigator and seem to know how to get to the bottom of things."

"I appreciate the vote of confidence."

"I know you will want to check on me, and that is fine. I am Mary Soderstrum. I am 39 years old and have a doctorate in Genetics and Master's degrees in Chinese and Biochemistry. I currently work at Fort Detrick. Do you know what that is?"

"Yes, it's a military base in Maryland that has been the home of bio-chemical weapons research for decades. They have a BSL-4 on the base. I also know it's highly secretive."

"I work on Level 4 of that laboratory and have Top Secret government clearance. What do you know about the current status of biochemical research into weapons?" asked Mary.

"Since Covid-19 and the conspiracy theories around it, I have done some reading, but there isn't a lot out there. I know of the various conventions and protocols many governments have signed stating they will do no such work. I also am aware that several countries have broken those."

"Good. I will ask you to play along with me for a bit, but I can prove everything I will be saying. Countries are allowed to do research around any known virus, bacteria, or chemical that could be used as a weapon. It makes sense. We want to know how to defend against the attacks of these agents. We look for vaccines and other ways of protecting the populace against them. Anthrax has been used as a weapon. It's been around for at least 2,500 years. Some feel it was one of the ten plagues described in the Bible. Long ago, they would infect sheep with Anthrax or find sheep that had died because of Anthrax and launch them with a catapult at the enemy – with some success. In Russia, in 1979, there was a leak from a microbiology lab that infected seventy-nine people. Sixty-eight died."

"You probably remember people putting Anthrax in envelopes and sending them to politicians and business people. If you catch Anthrax early, it is curable. We vaccinate sheep and sometimes cattle against Anthrax because that protects humans. The chances of getting Anthrax naturally, is very small."

"I assume you know about the Spanish Flu, the H1N1 virus that infected over 500 million people and killed over 50 million in 1918. We have learned a lot since then, which is why Covid-19 did not have the impact it could have."

"New technologies are changing how bio-weapons work. We have many new viruses and bacteria to play with. We can reproduce them at alarming rates, and we are close to being able to create our own virus."

"That isn't really possible, is it?"

"You ever see one of these?" Mary held up a 5.25 inch floppy disk.

I laughed. "Yes, that is a floppy disk. I haven't seen one in over 40 years. They still make those?"

"No. You probably remember the 3.5 disks. This holds 1.2MB. It is the high-density version. That is not even one good picture, and they are very slow. This is the latest thumb drive," she explained holding the thumb drive. It can contain one terabyte of data. That is roughly one million times more information in something smaller and faster."

"It is pretty remarkable the speed at which technology is growing."

"Do you remember Dr. Christian Barnard?"

"The first heart transplant."

"Excellent. Yes, in 1967. In 1964, they stated that would not be possible for at least 50 years. They are now performed regularly, along with liver and kidney transplants. We underestimate the pace of technology, and a great deal of technology is first developed behind closed doors."

"So, what is going on in your field?" I asked.

"At Fort Detrick, we are experimenting with gene-editing, using an advanced form of CRISPR."

"I've heard a little about CRISPR, but I don't know much." "If you decide to do this work, you will have to get to know a lot about it. For now, I will give you the barebones. CRISPR stands for Clustered Regularly Interspaced Short Palindromic Repeats. In essence, it is a gene-editing tool. A gene is a sequence of nucleotides in DNA or RNA that encodes the synthesis of a gene product, either RNA or protein. When genes manifest, the DNA is first copied into RNA. The RNA can be directly functional or can be the intermediate template for a protein that performs a function."

"Repetitive DNA sequences, called CRISPR, were observed in bacteria with "spacer" DNA sequences in between the repeats that exactly match viral sequences. It was then discovered that bacteria "write" these DNA elements to RNA upon viral infection. The RNA guides a nuclease (a protein that cleaves DNA) to the viral DNA to cut it, providing protection against the virus. The nucleases are named "Cas," for "CRISPR-associated. They are proteins. Cas9 has been a central player."

"It was in 2012 that researchers showed that RNAs could be constructed to guide a Cas nuclease (Cas9 was the first used) to any DNA sequence. The so-called guide RNA can also be made so that it will be specific to only that one sequence, improving the chances that the DNA will be cut at that site and nowhere else in the genome. Further testing revealed that the system works quite well in all types of cells, including human cells. I can see your eyes glossing over, so I will stop," Mary said, clearly excited about the subject.

"This is a lot of information, and it's quite complicated."

"I would not make a good teacher to beginners. My apologies. I'm sure this sounds like a textbook to you. What I have just described is like seeing the formula $E=MC^2$ and thinking you understand it. Those that deeply understand the theory of Special Relativity can be numbered in the tens of thousands. Special relativity applies to all physical phenomena in the absence of gravity. General relativity explains the law of gravitation and its relation to other forces of nature. It applies to the cosmological and astrophysical realm, including astronomy. With CRISPR, let's just say that we have gone from the floppy to the terabyte in ten years, not 40. So far, institutions and countries have stated they will not experiment with CRISPR on humans. You may have read about the Chinese twins who were created by CRISPR, a direct violation of the protocols. The government said they knew nothing about this work. I doubt that is true."

"So, you are working with CRISPR?"

"Everyone in the field of genetics is working with CRISPR. There is no doubt it will develop new strains of crops that will be drought tolerant and disease resistant, as well as need less water. Few have any doubt that we will be able to create vaccines more rapidly and potentially rid the world of some diseases."

"What is the problem?" I asked.

"My boss was asked three months ago to create a new virus. He was given instructions on what the virus would do to a person, how long it should take to inflict the damage, how it would be spread. He was also asked to create a vaccine to protect those who received it. The mortality rate was to be no lower than 80%."

"Your boss agreed?"

"At first, no. They told him if he didn't, he would never get another job in the field again, and if he told anyone, he would never see the outside of a jail cell again. He'd be in isolation for the rest of his life."

"Why not tell a Senator, the FBI, the CDC, or someone?"

"I have not seen it, but they showed him a file they'd created that has 'proof' he is a spy and a traitor. I suspect they have one on me as well. Of course, it is all lies. There is no one more patriotic than my boss."

"Why did he tell you?"

"I work in the lab. I would have found out. He wanted to give me the opportunity to get out before it all started. When I saw how fast this was coming together, I called you."

"How long before you have such a bug?"

"CRISPR is in its infancy. We have to find the right parts of the right virus. Then we have to find the protein that will do the cutting and pasting. After that, it must be tested. We estimate five to seven years at the earliest. My boss is more optimistic, and he is a visionary."

"You make it sound so easy, just start cutting and pasting."

"There is a lot of experimenting. We do not even know which genes do what. The Ebola virus has seven genes. The H1N1 has eight. Humans have 20-25,000, and all genes are of different lengths. The complexity is enormous. Without modern computers, CRISPR would not exist. We input massive amounts of data into the computer and then look for certain patterns. Then we experiment. It is only a matter of time before we figure out how to do what is being asked of us. Our budget is essentially unlimited. We would be naïve to think other countries are not doing the same. What if Russia, North Korea, China, Syria, Venezuela, or many other countries were to figure this out?"

"They have the resources to do this?"

"The actual process is not that complicated if you understand the mechanisms and have the resources. To some degree, it's trial and error. Computers can speed that process up. Let me give you some perspective. In 1951, the

Univac 1 could perform 2000 instructions every second. By 1980 we were at 6 MIPS at 6MHz. In 2002, the fastest was the ARM11, which computed 515 million. In 2020, the AMD 3990X could execute 150 billion per second. In that same year, El Capitan, a DOE/NNSA multi-processor computer, could execute over two quintillion calculations per second. That is faster than if every person on the planet were to do a calculation every second for eight years.   Things have gotten faster since then. Imagine putting that power to work around the area of genetics, for better or for worse. So yes, any country that is willing to spend the money can do gene-editing. If you understand genetics at a deep level, you can read journal articles that indicate several countries are well on their way."

The shivers returned to my spine. I was going to start having sleepless nights.

"And how do you think I can be of help?" I asked.

"I'm not sure you can, but I know you can raise questions that others would listen to."

"What information can you give me that would help?"

"I can give you questions to ask, people to talk with, and I can let you know how close we are getting."

"I don't even know where to start."

"I suggest going to the Senate tomorrow morning and listening in on the Armed Services Committee. My boss will be there. Please understand that none of the Senators probably know what we are doing. This work is as deep as it gets. I doubt the President knows."

"Who does know?"

"We have met with General Fleming that oversees the base. That's it. We report directly to him."

"What if he's a bad guy?"

"The thought has crossed both our minds. We have an encrypted email that goes directly to a private email of his.

He does not come to Level 4. We go to level 1 to meet with him."

"This is a lot to take in. I appreciate your sharing with me and trusting me. That means a lot. I will do my best not to let you down. What is it you want to happen?"

"To be honest, I don't know. What we are doing is against the law and is wrong, not to mention potentially very dangerous," said Mary.

I stood to leave and noticed another picture on the mantel. "Who is this? I assume the young woman in the lab coat is you. Who is that with you?"

"That is Dr. Clint Williams, my boss. He worked with me on my PhD. and gave me my first and only job. He is a world-renowned virologist and epidemiologist."

"I think I've heard of him. Is he full-time at the lab?"

"No, he works at several labs on different projects. He keeps them very separate because of the patents. When you work for the government, any "invention" you create on their time, is their patent. The government owns 11 patents of things he created. He has 17 patents on things he created outside the military. You may have seen him on TV during Covid-19. He does not like publicity."

"Have any of those patents produced any financial benefit?"

"I will let you do your homework, but let's just say he doesn't lose sleep over finances."

"Do you participate in any of that work outside the military?"

"Not yet, but expect to."

"Why does he continue to work for the military?"

"The military has access to things the private sector does not, including computing power, and secrecy. Until now, that has not been an issue."

"Remarkable that he can stay off the television screen."

"He prefers staying out of the limelight. He has no need to accept credit for the work he has done. In the world of viruses and bacteria, he is quite well known."

"Sounds like quite a guy."

"He is."

"Does he know you are seeing me?"

"Not yet, but he will. At some point, that is unavoidable."

"I know you are very loyal to him and the amazing work you all do. I honor that. My guess is you have been wrestling with this since the day it happened."

"Both of us have. We talk about it almost daily."

"I think with both your minds at work, you will find a solution."

"We might find a way out, but that is just us in this country. It would be unwise to think other scientists in other countries are not doing the same things we are. We have no control over them."

"Fair enough. I look forward to seeing him tomorrow. I almost forgot." Reaching into my pocket, I pulled out a cell phone. "This is yours. To reach me, speed dial the number 1. I have your number, and I will only call if I really need something. You do the same. No one else has these numbers. Keep this phone safe and not lying around. You know how to reach me via other means if necessary."

"Thanks."

We shook hands, and I left. I did my best thinking while walking. My cellphone recorded the thoughts and questions I came up with. The adrenaline rush had calmed down. Part of me was hoping there was already a virus created by humans. Once I came back to reality, I was glad there wasn't. I knew the clock was ticking, and something needed to be done.

4

I'd lived in New York for twenty years. My life kept me on the move most of the time, and New York was a good central point for most of the powerbrokers in the world. It was also an easy place to hide when the need arose.

My success had allowed me to move from a rent-controlled apartment in the village to a sixteenth-floor Co-op on the Upper West Side with a view of the river. The people at the door were an added sense of security, but there had been times I worried for their lives. If someone wanted me, they wouldn't stop at those wonderful folks.

Walking in NYC, one could feel the energy. It was palpable and, even at 3 am, it was there. Central Park was the great oasis. I have no doubt if it was covered with buildings, NYC would be a different place.

The greenery in the summers was breathtaking, especially with the tops of buildings popping above the tree line. I enjoyed sitting on the benches, watching the wide variety of people that passed by. Walkers, joggers, strollers, bikes, horses, and skateboards were all regular attendees to the runway that lay before me. I realized how innocent we all are. In the bowels of laboratories around the world, scientists were busy creating biochemical cures to diseases as well as new diseases that could wipe out the planet. Ignorance is bliss. I was no longer ignorant.

When I started a new investigation that had the potential of being dangerous, I told my editor, two of my best friends, and my mother. There had been a few times I would text them hourly with my location. If they didn't hear by the next hour – 24/7 – they were to call in reinforcements. They were all anxiously supportive. My mother didn't understand why I did what I did since I was already financially secure. All I could tell her was that there was a deep and burning need in me to make things right. People who understand the Enneagram would see that I am a very strong type one. My wing is a two – I try to be a caring person. I hope a healthy one. I use the Enneagram, a personality system, constantly sizing people up, how healthy they were, and were they becoming more integrated or going the other direction. Most of this is done subconsciously, but it comes in handy. I'd met Don Riso, one of those who created the more psychological variation of the Enneagram. For those who do not know about it, I encourage you to spend some time investigating the system.

I'd handed out phones to people in all my investigations. When I got to know them and what type they were, it helped me know how to approach them. It was far too early to call these people who'd all said they owed me a favor, but the thought had entered my mind. In all, I think about 15 people around the world has one of those phones from previous investigations.

Mary was a 5/6 on the Enneagram. She is a brilliant researcher and very loyal – until being pushed off the edge. I could tell in our conversation that she was uncomfortable. Not with seeing me, but with an internal voice telling her she was not loyal to her boss or the country. I was grateful that voice was more loyal to the totality of humanity. I tried to speak to her loyalty and her rational mind. At this point, I would say it had worked.

That night, I went out to dinner, which I did about four nights a week. I was not the world's greatest cook, although I did have a few specialties. The train for D.C. left at 7:00 pm from Penn Station. It was an express and would get me into D.C. about 9:00. I decided on a deli sandwich from a place just outside the terminal.

When we were underway, I got my computer out and started researching Dr. Clint Williams. He was an impressive figure. He'd been in the field of genetics for decades. Being part of CRISPR from the outset in the private sector was where I was looking for a few of his patents. I tried to find any big money items. Then I found it. He was a key player in the main genetic testing for ancestral information. He'd made tens of millions of dollars, and I guessed he received residual income. Mary was right; he was not losing sleep over finances.

The meeting in the morning started at 8 am, in the Senate building. The Armed Services Committee had several sub-committees, of which Emerging Threats and Capabilities (ETC) was one. They would make recommendations to the full committee that would move legislation forward. All bills had to have sponsors. The session I would be attending was not a hearing on a bill but a means of checking in with those with information about emerging threats and capabilities. Since Covid-19, this committee had become more active. Members of the CIA, FBI, military, and the ODNI (Office of the Director of National Intelligence) were always asked to be there. On rare occasions, a President might not allow someone to testify. When that happened, a political storm usually ensued. The lines between the three branches of government were currently still not that clear. The early 21st century had seen those boundaries pushed and blurred. More time was spent in federal courts with different parts of the government trying to stop the excesses of power by the others than just about anything else.

Union Station was four blocks from the hotel where I like to stay, and that was three blocks from the Senate Building. I grabbed a bit more to eat on my way to the hotel and, once I got there, I read a little more and went to bed. It had been a long day, and I didn't really feel up to meeting with friends, which I often did when I was in town. I would be spending a great deal of time in D.C. over the next several months.

5

Many committees and sub-committees had two types of meetings. There were those open to the public where people like me could sit and watch. We were not allowed to ask questions. Most of the time, these meetings were well scripted. Everyone knew beforehand the questions that would be asked and the answers that would be given. Journalists always had more questions and tried to get those answered as Senators walked back to their offices. They rarely got what they wanted. Good politicians knew how to answer a question without answering the question.

The second type of meeting were closed-door meetings. These are where specific items were discussed that were classified in nature. The one I would attend was open. I expected those being questioned would have prepared statements, and then they would go into executive session.

As a journalist, that is where contacts made the difference. The world runs on leaks. If everyone knew how to keep their mouth shut, a great deal of journalism would perish. Fortunately for everyone, humans like to chat. The person who is the source of the leak will often go on national television angry about the person who created the leak.

Today was a scorcher in D.C. I was dripping by the time I got to the Dirksen Senate Office Building. The humidity was almost unbearable. Normally, I go for a run early in the morning. That habit tends to slide when I start a new project.

Finding your way around all the government office buildings can be difficult. First, you have to get in. A press card helps with that. There are few maps of any of the buildings, probably because they want terrorists to have to do some work and not give them the complete layout of every building online. I'd been to many hearings in the building, so I knew where to go. Even the early meetings started late. I grabbed a cup of coffee in the cafeteria and continued my work on Williams and Soderstrum.

The picture of Williams made him look about 55. His birthdate was listed as 1956, so he was in his 70s. Dr. Williams did not like to have his picture taken. He was a medical doctor with a PhD. in Neurobiology. He had master's degrees in Russian, Computer Programming, and Music (in piano). Nowhere could I find his estimated net worth. I did find a list of patents, mostly names of things I'd never heard of other than the ones I'd already read about. I would research that later. He was listed as the Director of the Fort Detrick BSL-4 lab. None of his other positions were mentioned.

I walked into the committee room, which sat about sixty visitors. On the dais were seats for about fifteen Senators with some folding chairs behind them for aides. Two long tables were between us and the Senators where those giving testimony would sit. Currently, at 7:55, there were five senators and three speakers. I took a seat near the aisle.

Senator Kathy Brower from New Mexico was on the dais. I'd known her for years. She looked up, saw me, and walked over.

"Hello Tom, what brings you here? We haven't done anything wrong, have we?"

"Good morning Senator. Of course not, you are the best." "So, why are you here?"

This was now a common practice. Many people knew who I was and what I did. They got nervous just seeing me. That was both a plus and a minus. I could no longer be incognito. They were all on guard.

"As you know better than I, the world of virology is changing. Since Covid-19, the recent resurgence of Ebola, the increase in cases of malaria and its spreading, and smallpox, I've taken an interest in what is going on and why. I could think of no better place to come than here."

"I am not sure you will learn anything here today, but I am glad you are attending."

"Thank you, Senator. I hope you and the family are well."

"Yes, we are. Thank you for asking. Now I'd better get on with the meeting."

"Good luck."

As the Senator was walking back to her seat, General Fleming walked in. He liked making an entrance and tried to be the last one seated at all meetings. Generals had egos too. This was the general from Fort Detrick, where Mary and Clint worked.

She swung the gavel. "I call this session to order. We are here today to listen to updates on the state of biological and chemical weapons. I will ask each of you to read your statement, and then we will open it for questions from the senators." Eight senators sat on the dais. There were several empty seats. Those absent did not think this would be an important meeting. That was not an uncommon practice.

Everyone pulled out their statements to be read. It was a formality for the public. The real work, if there was any, was done in executive sessions and between sessions in the back rooms. I was here to get a better sense of who the players were.

Senator Brower opened the hearing. "Before I begin, I would like to read from a statement given by Senator Nunn on July 23, 2001. I will enter this into the record but wanted to read a portion of it. I apologize for its length but feel it is important and potentially prophetic."

Senator Nunn: "Mr. Chairman and members of the Committee: Thank you for the opportunity to testify today on the threat of biological weapons. Two years ago, Mr. Chairman, presiding over a hearing of this same committee on this same subject, you asked: "Are we prepared?" The answer then was no. Your efforts and the efforts of others since then are forcing us to find a better answer—and I thank you for your persistent emphasis of this issue.

It was challenging to play the part of the president in the exercise Dark Winter described by Secretary Hamre. You often don't know what you don't know until you've been tested. And it's a lucky thing for the United States that—as the emergency broadcast network used to say: "this is just a test." It is not a real emergency. But, Mr. Chairman, our lack of preparation is a real emergency.

During my 24 years on the Senate Armed Services Committee, I've seen scenarios and satellite photos and Pentagon plans for most any category of threat you can imagine. But a biological weapons attack on the United States fits no existing category of security threats. Psychologist Abraham Maslow once wrote: "When all you have is a hammer, everything starts to look like a nail." This is not a nail; it's different from other security threats; and to fight it, we need more tools than the ones we've been using.

Our exercise involved a release of smallpox. Experts today believe that a single case of smallpox anywhere in the world would constitute a global medical emergency. As members of this committee know, a wave of smallpox was touched off in Yugoslavia in 1972 by a single infected individual. The epidemic was stopped in its fourth wave by quarantines, aggressive police and military measures, and 18 million emergency vaccinations to protect a population of 21 million that was already highly vaccinated.

Mr. Chairman, we have effectively only 12 million doses of vaccine in America to protect a population of 275 million that is not highly vaccinated and is therefore highly vulnerable. The Yugoslavia crisis mushroomed from one case; our situation began with 20 confirmed cases in Oklahoma City, 30 suspected cases spread out in Oklahoma, Georgia, and Pennsylvania, and countless more cases of individuals who were infected but didn't know it. We did not know the time, place or size of the release, so we had no way of judging the magnitude of the crisis. All we knew was that we had a big problem and a small range of responses. One certainty was that it would get worse before it would get better. As you know, Mr. Chairman, effective smallpox containment requires isolating those who are sick and vaccinating those who have been exposed. Isolation is difficult when you're not sure who has it; vaccination cannot stop the spread if you don't have enough of it.

Many participants in the exercise would have been much more in their element if we had been dealing with a terrorist bomb attack. The effects of a bomb are bounded in time and place. After the explosion, the nation's leadership knows if you're injured and the extent of the damage. We can begin rebuilding. Smallpox, on the other hand, is a silent, ongoing, invisible attack. It is highly contagious, and spreads in a flash—each smallpox victim can infect ten to twenty others. Because it incubates for two weeks—it comes in waves.

The most insidious effect of a biological weapons attack is that it can turn Americans against Americans. Once smallpox is released, it is not the terrorists anymore who are the threat; your neighbors and family members can become the threat, and can even become the enemy, without strong and effective leadership at every level of government including health officials. The scene could match the horror of the Biblical description in Zechariah (8:10): "Neither was there any peace to him that went out or came in because of the affliction: for I set all men every one against his neighbor."

At the same time, a biological weapons attack cuts across categories and mocks old strategies. For more than two thousand years the first rule of war has been to know your enemy. In military language, this means that when you face a battlefield scenario, you draw up an order of battle—you estimate the number of tanks and planes and troops of the enemy, their intelligence capabilities and other resources. But in this case, the order of battle is our own people, traveling, engaging in commerce, and spreading the disease. And there are few reliable numbers—you don't know who initially released it, how much more they have, or where they are. And the usual responses to an attack are impossible: "Engage the enemy; open fire; stop their advance; bring out the wounded." You can hardly know who is wounded.

For the participants, this exercise was filled with many such unhappy discoveries and unpleasant insights. Number one: We have a fragmented and under-funded public health system—at the local, state, and federal level—that does not allow us to effectively detect and track disease outbreaks in real time. Two: Since the disease has not been seen in the United States since 1949, very few health care professionals recognize the smallpox virus, so initial cases could be sent back home infectious, even after appearing at doctor's offices and emergency rooms. Three: Lab facilities needed to diagnose the disease are inadequate and out of date. Four: There is insufficient partnership of communication across federal agencies and among local, state, and federal governments. Five: The only way to deal with smallpox is with isolation and vaccination, but we don't have enough vaccines, and we don't have enough room, resources, or information for effective isolation. Six: A biological weapons attack will be a local event with national implications, and that guarantees tension between local, state and national interest. In our exercise, the governor of

Oklahoma asked for vaccine for every one of his citizens—
as he had to in the interests of his state. The president said
no, as he had to in the interests of the nation. Naturally, this
will demand a high degree of coordination, because of the
diverging interests, and because key players and partners are
answerable to different leaders. Seven: Hospitals run at
capacity all the time: a surge in patients from smallpox,
combined with the inevitable infections of hospital
personnel, and the flight of some fearful health care
professionals, would create a catastrophic overload. Eight:
There will be a dearth of information on this kind of event.
My staff and cabinet could not tell me ten percent of what I
wanted to know: "How many cases are there right now?
How many more are coming? When and where did the first
infections take place? Who released it? What's the worst
case scenario?"

And there are many tradeoffs. One of the biggest: We
have 12 million vaccines; that's enough for <u>one</u> out of every
23 Americans. Who do we decide to vaccinate?

Other tradeoffs are do you take power from the
governors and federalize the National Guard? Do you seize
hotels to convert them to hospitals? Do you close borders
and block all travel? What level of force do you use to keep
someone sick with smallpox in isolation? Do you keep
people known or thought to be exposed quarantined in their
homes? Do you guarantee 2.5 million doses of vaccine to the
military; or do you first cover all health care providers? Do
you take strong measures that may protect health, but could
undermine public support or destroy the economy?

And finally: How do you talk to the public in a way that
is candid, yet prevents panic—knowing that panic itself can
be a weapon of mass destruction?"

My staff had two responses: "We don't know," and
"You're late for your press conference." I told people in the
exercise: "I would never go before the press with this little
information, and Governor Keating—who knows about

dealing with disaster, said: "You have no choice." And I went, even though I did not have answers for the questions I knew I would face: "How bad is it?" "What's the plan?" And "Why, after all this time, isn't there enough smallpox vaccine?"

Naturally, there are some skeptics anytime you describe a dire threat to the United States. I want to tell the Committee: I am convinced the threat of a biological weapons attack on the United States is very real. As Secretary Rumsfeld said in his confirmation hearings: "I would rank bioterrorism quite high in terms of threats ... It does not take a genius to create agents that are enormously powerful, and they can be done in mobile facilities, in small facilities." An experiment some years ago, showed that a scientist whose specialty was in another field was able to weaponize anthrax on his first attempt for less than $250,000.

Hundreds of labs and repositories around the world sell biological agents for legitimate research—and the same substances used in legitimate research can be turned into weapons research. In addition, the massive biological weapons program of the former Soviet Union remains a threat, to the extent that materials and knowhow could flow to hostile forces. At its peak, the program employed 70,000 scientists and technicians, and made twenty tons of smallpox. One Russian official was quoted some years ago in the New Yorker saying: "There were plenty of opportunities for staff members to walk away with an ampule." (end of passage)

"The Senator then goes on to say what we had done to that point and twelve things that still needed to be done. We have yet to accomplish most of those steps. He ends his remarks with the following," stated Senator Brower

"According to some historical accounts, what pulled America back from financial panic in March of 1933 were three things President Roosevelt did immediately on taking office: he ordered the banks to close temporarily, he proposed emergency banking legislation, and he explained his plan to the public in the first of his regular national radio broadcasts.

If he had not talked reassuringly to the American people, his plan might not have worked. But if he had talked, but had no plan, his talk would not have been reassuring. In the event of a biological weapons attack, no president, no matter how great his natural gifts, will be able to reassure the public and prevent panic unless we are better prepared than we are right now. If we are well prepared—with the ability to detect the disease quickly, report it swiftly, and isolate and vaccinate all those who came in contact with it—then the president of the United States will address the American people with courage and confidence, and the people will respond in kind. How the president is able to address the public on that day will depend in large part on how we all address this issue today. Thank you."

"That is the end of his testimony. All of you are here today to let us know how our preparedness is coming along, and what we know about foreign work being done on the creation of these types of weapons. Senator Nunn's remarks were solely about smallpox, a known virus. We are more interested in the unknown viruses. We will start with Ms. Wilson from the CIA, followed by Mr. Vladstock from the FBI, then Mr. Spencer from the NSA (National Security Agency), Dr. Williams from Fort Detrick, and finally, General Fleming, who oversees chemical and biological weapons for the military. As always, we apologize to the public in attendance. Much of the information on this subject is classified and cannot be mentioned in these hearings. We will take a short break after the questioning period to determine if we will need to call for an executive session. Ms. Wilson," said Senator Brower.

We'd been given copies of their statements, which were five to ten minutes in length. It was evident that most of what was important to say couldn't be said. The CIA, FBI, and NSA all stated they were concerned about what their contacts were telling them about the progress being made in a variety of countries. Each of the departments believed that the actual creation of a biological or chemical weapon of mass destruction was still at least ten years away. They did voice concerns that it was a matter of when not if. None of them gave very specific ideas about what could be done about it. Our research into these types of weapons was based on a defensive posture. How can you research something that doesn't exist?

"Dr. Williams, can you give us your perspective looking from the inside?" asked Senator Brower. Up till Dr. Williams, the speakers had been bureaucrats, not scientists. He took the microphone.

"Thank you, Senator, for inviting me here today. As you know, I work on Level 4 of the laboratory at Fort Detrick. It is the oldest Level 4 lab we have. In case you are worried, it has been updated since the 40s." Everyone laughed.

"We have entered a new age in biological weaponry. I will not discuss chemical weapons because 1) I don't work with them, and 2) while their impact is immediate and gruesome, it pales in comparison to what a biological weapon could do. The change is happening for two reasons. The first is computing power. We are finding cures and vaccines to diseases much faster now because of this speed. Artificial Intelligence is taking some of the guesswork out of the process. What used to take years has been cut down to months in some cases. The other contributing factor is what the world knows as CRISPR, a gene-editing system that allows us to change DNA and RNA. Both of these fields are immensely complicated. If you read the research from around the world, you get glimpses of how fast we are becoming able to change and potentially create new life forms. I am confident that within ten years, a virulent form of weapon will be created that could decimate the world."

"Pardon my interrupting," said a senator from West Virginia. "You are saying that a country will be able to weaponize a virus within ten years?"

"Yes, Senator, that is my belief."

"I am puzzled by that doctor. We have protocols and conventions in place that outlaw that research and my understanding is that all Level 4 labs are required to have inspections to specifically look for this type of work."

"You are correct, Senator. Let me ask you a question if I may."

"Certainly."

"If I were to tell you that three people were coming to your house to look for baking soda, and if they found some, they would arrest you and burn your house down, and you had baking soda in your house that you needed, what would you do?"

"I would hide it somewhere or take it out of the house while they were in it."

"Of course you would. There are no surprise visits to laboratories, and we only go to Level 4. Let me ask another question. Let's assume you can't take the baking soda out of the kitchen area. What would you do?"

"I would put it in a different container and label it something else, like flour."

"Precisely. You can't look at 300 vials of biological entities and know by sight what is what. All labs have detailed tracking, but those are easily forged."

"What about the equipment they use? Wouldn't you see they are using CRISPR?" asked another senator.

"Yes. All labs use CRISPR. At Fort Detrick, we are trying to find better responses to smallpox, ebola, and several others. CRISPR expedites that research. We need CRISPR for defensive purposes, but this does not mean nations can't use it for offensive ones."

"And what do you suggest we do?"

"The world is at a crossroads. Unless you can find a way to truly find and examine all Level 4 labs on a very deep level, you will not succeed. I think we all know there are countries that do not care about life. You do not need a Level 4 lab to carry out this work. It can be done in your basement. The lower the level of sanitation, the higher the potential for a leak. My biggest concern is that a virus is created and leaks out before a vaccine can be produced."

"So, we would just find a vaccine."

"You are aware, Senator, that there is no vaccine for AIDS. We have treatments, but no vaccine. Some viruses are highly resistant to vaccines. I have no doubt that those countries trying to create a virus are doing so with the idea that the vaccine is very complicated. The spread of the virus could be so fast and so effective that up to 50% of the world's population would die before it would be found."

"Then why would any country do this?"

"Imagine Senators, that you have a weapon that will destroy people at will. You vaccinate your population, then you find an isolated population and infect it. The world watches them die. You then hold the world hostage. Money, power, or both. What will you do? Will you let your citizens die, along with yourselves? Will you turn over power? Will you hand them 5 trillion dollars? This is going to happen. I do not know when, but it will happen," said Dr. Williams.

"We are not engaged in such weaponry, are we?" asked Senator Brower.

"No," he said, gazing over at the General.

"Do you know what countries are involved in this type of work?" asked another senator.

"My guess is any that can. Think of the power you would have if you had such a weapon. Even a third world country could hold the world hostage. It is a question of resources and finances. This is not high school biology or computer programming. There are not that many people on the planet that could put the pieces together. There are perhaps ten countries that could do this now."

"Thank you, Dr. Williams for sharing. We may have some other questions for you if we go to executive session."

"I would like to say one more thing."

"Please do."

"I have been honored to work at Fort Detrick for the past four decades. Most of that time has been full or half time. As you know, I also work with the private sector and teach. The work I do in each sector is unique, and there is no cross-fertilization. We have accomplished a great deal for which I am proud. I am now in my 70s and feel it is time to pass the torch. I am using this forum to announce my semi-retirement. My hope is to work quarter-time if the military will have me. I will sit in your executive session if you wish, but that will be the last time you will see me in this setting." There was a gasp in the room and an icicle glare from the General.

"We are all saddened to hear that Doctor. We value your work and your voice. We wish you well in whatever endeavors you choose take on. I hope there is some fishing or something relaxing in the mix."

"Of that, I have no doubt."

"General Fleming?"

"Thank you, Senator. I did not know about his retirement, and am a bit shocked. He is a one of a kind specialist. I am not sure we will find people to replace him. It may take ten. His knowledge and expertise are unsurpassed in the world. We will miss him. I do look forward to having him with us quarter-time for the next couple of years. He does more in one hour than most do in a day. Now, to my statement. I will be brief since I believe we will be going into executive session. Yes, other countries are doing what the doctor suggests. Russia, China, Iraq, and Venezuela are at the top of the list. This is no surprise. I will give more details on how we know this in the private session. As the doctor said, there is no counter-offensive for a weapon you do not know. We could start to do similar work in our laboratories, but, as of now, that is not allowed."

"So, we all have viruses. Do we just all unleash them? Weaponize them? Then the world dies," said Dr. Williams.

"You may remember we had and still have MAD, mutually assured destruction. If a nation decides to launch a nuclear weapon, they resist because they know the country about to be destroyed will launch their missiles and destroy their country. We could do the same. The chances are slim that a country with a virus would use it if they knew it was coming back at them. If we cannot make a credible threat that we have a virus, then MAD doesn't exist," the General stated.

"There is a big difference with viruses. With nuclear weapons, you know where they are coming from. We know where the submarines, silos, and airplanes are at all times. You don't launch viruses in the same way. Depending on the virus, one person could distribute a virus that would infect and kill millions. How would you know who you counterstrike at?"

"We have our methods."

"Like the same methods that stated Hussein had weapons of mass destruction?"

"That was different," the General stated, with nods from Mr. Spencer and Ms. Wilson.

"I say with great respect, none of you have a clue as to what is going on and what you are up against. It may unfold well before you have figured out a strategy. Let's play this out for a second. Pretend you have sources that let you know a virus has been created at a laboratory in China. You know this for a "fact," although we know how good facts are. You and the President determine this is reality. You make threats to their government. They invite you to come and inspect the facility. As I said earlier, the virus has been moved. You can't spot the movement of viruses or bacteria on a satellite like you can missiles. Because of this, you decide to just destroy the facility, and you send a drone and wipe it off the map. Do you actually believe a nation is going to stand by and let you do that? Perhaps Iraq, Syria, or Venezuela, but certainly not Russia or China. What will you do?"

"We have plans," stated the General, clearly getting agitated.

"Of course you do," said the doctor.

My guess was the semi-retirement would come as a surprise to Mary as well. I wondered if she would quit or be promoted. Perhaps after this hearing, the government would give their approval to enter the virus arms race. This part of the hearing ended after a few more questions were asked of the panel. There would clearly be an executive session. I needed to keep tabs on that committee because my hunch was they would be very busy over the next year.

The main character in my story was about to leave the building. I would be interested to see if they would let Clint continue to work part-time. They have probably already started looking for a replacement. I guess they didn't need Dr. Williams for the closed session.

In my line of work, you need to do things that are less than above board from time to time. I always carried a tablet, camera, plenty of money, a change or two of clothes, two other phones, and some tracking devices. I started every day with a clean tablet and phones. I never put any of my work in the cloud because I knew lots of people had access to it, no matter what they tell you. I used a special VPN a security specialist told me about to make it very difficult to trace me. The actual writing I did on a computer that was not attached to the web. Every day I would save what I'd compiled on three different flash drives. If someone decided to burn the house down, which they had in the past, I didn't want to lose the work. The more paranoid I became, the more security I added around the house. Normally, the last few weeks of writing would be done in isolation, far away from everyone.

Senator Brower came over and had a chat with the doctor. When I saw the doctor putting his things together to leave, I left the building first. We all exited by the same door. I was trusting he was not paranoid at this time. As he walked out of the building, a couple of reporters asked him questions. He was evasive and blunt and walked away. I followed him to his car. The one-year-old BMW sat amidst Subarus and Toyotas in the parking lot.

"Dr. Williams, thank you for your candor at the hearing. I am sorry you are retiring, but appreciate what you have done and what you will do."

"Thank you. And you are?"

"Tom Lassiter."

"OK, take care Tom. I'm not fully retiring, just changing focus."

I never used my real last name, and Lassiter was the normal fake I used when first meeting people. He didn't recognize me. I was more known by my words and name than by my face. Rarely did I give interviews. I'd been asked by Oprah, CNN, 60 Minutes, and others. I declined. He was quite distracted, so he didn't notice me drop a device into his leather satchel. He was not the most organized person on the planet.

As I walked behind his car, I attached another of my miniature tracking devices. As he pulled out, I turned on my tablet and watched him drive away. The range of the car tracker was four hundred miles, and the one in the satchel, two hundred miles. Having no idea where he lived, what he was thinking or feeling, or where he was going, I decided to follow him to his destination. I called the rental car company and got a car. I caught up with him after about forty minutes and stayed a couple miles back.

I enjoyed all types of music. Pat Metheny, the Beatles, Windham Hill, Joni Mitchell, some Christian music, Chopin, and Led Zeppelin were at the top of the list. After an hour, Clint pulled off to get gas and something to eat. I had plenty of gas and didn't want to be seen, so I stayed in the car.

Clint returned and started to drive out of the station, then, for some reason, pulled back in and parked. I happened to notice a red sedan had gassed up close to him. It was a military vehicle. The sedan followed Clint out and then followed him back in, parking on the other side of the station. When the doctor left, the car followed. The military wanted to know what he was up to. I doubt he was aware of that fact.

6

When I write, I try to get to know the people I write about. When possible, on a first-name basis – Clint.

I was careful as we drove not to let the man in the red car see me. We were on 95, and Clint showed no signs of getting off. Because of the range of my tracking device, I pulled off for gas and food at different times than he and the red car. The military person would have to gas up and eat at the same time Clint did unless they had a tracking device, which I doubted.

Over seven hours later, we pulled into Boston. I followed him more closely to Newland Street on the south end of town. So did the red sedan. The doctor parked outside a townhouse and let himself in. I wrote down the address. I was paying more attention to the sedan that had parked down the street. After a few minutes, the driver got out, walked to Clint's car, and put a tracking device on it. Then he drove away. I didn't think that was really fair, so I got out of my car, took the device off Clint's car, and put it on a different one. The military wanted to know what he was up to. I recognized this style, and it had a cellular connection so they could monitor it from anywhere. My guess was the man in the car was gone. The device would relay back to base where he went.

I figured he was heading for the National Emerging Infectious Disease Laboratory located on the Boston University Medical Campus a few blocks away. It was a Level 4 facility and was a five-minute walk from where he was staying. My tablet told me he'd owned the townhouse for six years. It was 10:00, and I decided he was probably going to bed. The hearing and the drive had been long. I found a hotel nearby and checked in. My tracking device would beep on my laptop when the car moved, so I wasn't worried about him going anywhere. I activated the one in the satchel. It showed up on the first floor of the three-story townhouse. I don't underestimate the paranoia of the military and their ability to track, so I decided to use another device to search for tracking devices on my car, person, and in my residences. They could make life difficult.

Clint spent the early morning in the lab, went back to his townhouse, and then left. If he was returning to D.C., I was puzzled why he'd driven. He could have saved hours taking a plane. We followed the same routine we had driving north, but when we passed Philadelphia, we went south to Delaware. Cars thinned out once we went through Dover. We were hugging the coast, a couple miles inland. I stayed well behind him. When he turned left, I stopped. There was a signpost with a number on it, but no mailbox. Everyone had to have some way of identifying their address in case of an emergency. I searched the records and discovered he owned a house here with ten acres of property. It was not purchased under his name, but that of a company, "Farm and Sea Supply Co." Clint was full of surprises. It was a short walk to the beach, and he had no neighbors. Perhaps this is where he got away from it all. He was about two hours from Fort Detrick.

I wasn't going to knock on his door, but I was curious. I took my chances and left. It was 6:00 pm, so I thought of going to the Jersey Shore to stay at a beachfront hotel and get in a night and morning swim. It was still light, and I am a bit of a homebody. I headed back to New York.

What the next steps were was unclear to me. I could confront Clint or the General. That would be met with denial, and I had no evidence. I could leak something to Senator Brower or numerous other political figures. They could hold a hearing that would be met with denial as well. My gut told me Clint was a good person. From his testimony, he deeply believed several countries were involved in this biological weapons race. I am confident he knew conventions and protocols did not stop most, if any, countries. I suspected he'd been working on a strategy for some time. I simply needed to figure it out.

Going through my contact list, I found one who might be able to help. The TV show, Who Wants to be a Millionnaire, had the phone-a-friend idea. I was about to phone a friend. I gave him a call from my phone in the car.

Charles Johnston is a professor at UC Berkeley in Epidemiology. I'd met him while doing an article on Covid-19. He was brilliant and well respected.

"Hello, Charles."

"Charles, this is Tom Armstrong, the reporter."

"Yes Tom, how are you?"

"I am fine, thank you, and you?"

"Busy as always, but well. What can I do for you?"

"I had a scenario for you that you might be able to help me with."

"Glad to do what I can. I only have about fifteen minutes right now."

"I will plant the seed, and you can get back to me if you come up with something."

"Fair enough," Charles replied.

"Let me give you a few assumptions first. A biological weapon is being created by several countries, potentially including our own. A scientist wants to stop the madness. How could they do that? I will assume no country will let others into their laboratories where the work is taking place. I assume, if a country really wanted, they could hide these labs anywhere, and they would not have to be Level 4 labs if they didn't care about human life. What might this person be able to do?"

"Wow, you do come up with them. Mind me asking what brought this on?"

"It started with people stating COVID-19 was a created virus. That got squashed, but with all the gene-editing going on, it's certainly on the horizon."

"So, being the investigator that you are, I should not assume you know that someone is actually doing this."

"Correct."

"All of your assumptions are correct. This has been discussed at meetings and over drinks. I have not put a lot of thought into the matter, but I will give it some thought and get back to you if I come up with something that might actually work."

"One final question for now, what would someone need at this time to create such a virus or bacteria within two years?"

Charles laughed. "I don't think that is remotely possible. However, if they had virtually unlimited resources, they'd need a Level 4 lab. They could do it with less if they didn't care about safety. They would need the highest and most modern CRISPR technology, a machine that could do thousands of a variety of tests (that doesn't exist at this time), and massive computing power to pour through the data. On top of that, they would need a great deal of luck."

"Sounds highly improbable."

"At this time, yes. We just don't have all the technology needed to do the work effectively. The process is time-consuming. It's one of the reasons it takes so long to discover vaccines."

"Thank you, professor. I won't keep you. You have my number. Give me a call if you come up with something."

"I will do that."

Looking into things takes time. We have this notion that crimes are solved overnight and that when we see investigative reporting on TV, that the report took a day or two to create. Patience is a core virtue in this line of work. While waiting was not enjoyable, it was necessary.

I got back to my apartment, looked through the mail, checked the news, and all the phones. No messages from anyone. It was only eight, and I had a lot of energy. I decided to go out to dinner and called a friend. We ate in the Village and listened to some live blues. Then we hopped over to a comedy venue and laughed our heads off for two hours. The best medicine for the ideas circumnavigating my brain. I got to bed about 1:00 am and was sleeping soundly.

7

The phone rang at 6 am. Everyone knew I stayed up till 1 am and got out of bed around 9 am.

"Hello?"

"Hi Tom, Charles here. Oh damn, it's six there. I'm sorry, should I call back?"

"No, you are fine. What's on your mind."

"Your scenario got me to thinking. I ended up having drinks with some colleagues and threw the idea at them. We could only come up with one idea. We had several, but they all had significant issues we couldn't resolve."

I sat up, energized. "So, what's the idea?"

"So we had to make a few assumptions. We already talked about the technological parts. What we didn't talk about were the human parts. We will assume this person or persons will have a very deep understanding of advanced CRISPR technology and the means to advance it further. We will also assume at least one of the people is an expert in Deep Learning. They will have to create an algorithm to find the necessary data, compile it, and separate the wheat from the chaff. Then they unleash it in an isolated area. The world will figure out that it was a human-created virus. That should change things."

"That's great Charles, thanks for your efforts. That helps a lot. By the way, what is Deep Learning?"

Charles laughed loudly. "I will let your investigative mind look it up. It's part of Artificial Intelligence. I know you don't go on wild goose chases. Let me know what this is about when you can."

"You will be one of the first."

"Go back to bed."

"Thanks. Bye." Going back to sleep was not going to happen. I didn't think his idea had much credibility, but it did point me in a different direction. Clint would not only need a great deal of computing power, which he had but would also need a programming wiz to help with what he needed to compute. Now, I not only needed to learn about the world of viruses and gene-editing but of high-level computer programming. Both were well above my paygrade.

I'd be doing research in the field of AI. First, I needed to find out what it was. Like bacteria and viruses, there were a lot of angles and fields. I sat down at the computer and dove in.

An algorithm is a set of instructions computers use to solve problems. A computer program is a description of an algorithm in a form that both people and computers can understand; programs are written 'programming languages' with names like Python or C++. Imagine that you wanted to tell a computer-controlled robot how to make a quiche. You would describe a quiche recipe using a programming language, yielding a computer program. You would then give the robot the program and the ingredients for the quiche, and out would come the quiche. The quiche recipe is a kind of algorithm. The ingredients are inputs, kind of like data supplied to the system.

Algorithms were invented in the 9th century. Electronic digital computers allowed for more sophisticated algorithms. Weather modeling, facial recognition, and others are becoming more accurate as ever more powerful computers can deal with more complex algorithms.

The term AI was coined in 1956. By 1959, a computer could play checkers. The moves were made not by thinking but by learning. By looking at thousands of other games, it could statistically evaluate what the best next move would be. One might call this thinking, but the computer can't make a move outside the parameters the data gives it.

When a computer is programmed to beat the world champion chess or go player, they can't stop and plot a rocket's trajectory to the moon. In some ways, humans are no different. I can play chess, but I don't know how to plot a rocket's trajectory. I don't have the data or the know how.

Most people talk about AI on three levels. AI is seen as the broad umbrella. Artificial Intelligence is the notion that a computer can imitate human thought. The most elementary level is called Narrow or Weak (ANI)- what we have today. This type of AI can do repetitive functions but much faster than humans.

Siri, Alexa, Bixby, and others are ANI. They have been taught voice recognition. Try asking them an abstract question like "Why am I sad today?" "What is the meaning of life?" doesn't work. Unless a programmer has specifically put in an answer, which Siri will give every time, it usually says something vague. Computers don't think — yet.

AGI, or Artificial General or Strong Intelligence, is the ability to think like a human. This means the computer will be able to reason, solve problems, plan, learn, be innovative, imaginative, and creative, and make decisions in the midst of uncertainty. Most people believe we are at least twenty years away from that, even at the most basic level.

ASI, or Artificial Super Intelligence, is the ability of a machine to think faster, deeper, and broader than humans. AGI and ASI are what movies are made of. Robots that think one step ahead of humans and yet understand emotions. Some, like Elon Musk, believe ASI is dangerous. Others, like Ray Kurzweil believe ASI will benefit humanity. He sees a time when we will be able to plug our brains into the computer and learn things far beyond our current comprehension.

For now, we have ANI and are dependent on speed, amounts of data, and the ability to write algorithms that can process that data in meaningful ways. They do that in a number of ways, but there are two main ones.

Machine Learning is a strategy that teaches computers how to learn. The computer improves its accuracy through experience. It looks for relationships between data: correlations. By using standard deviation, normal distribution, the Bayes's Theorem, and feature extraction, among others, they learn.

This learning process takes place on three levels. The first is Supervised where the computer is given data and told what the outputs should look like. The computer searches the data and does so quickly. Unsupervised learning is when the computer is given large amounts of data and told what the end result should be, but the computer figures out the algorithms to get there. As the computer tries and fails, it learns and adjusts the algorithms. It is still based on the input data. The final style is reinforcement learning. Here, like an autonomous vehicle, the computer has input but also can take in real-time data. It creates new algorithms to fit the new data to drive safely. That new data is then in the memory bank for future similar situations.

The next level is called Deep Learning. With the size of data banks and the processing speed of computers, algorithms are created to look in many different layers or levels of data for patterns. This has increased the ability to do voice and facial recognition, as well as to speed up the ability to find and test new medications. The computer looks at the chemicals in the medication, then compares that to brain chemistry and the impact of other medications with the same or similar chemicals. They can predict where and how a medication will impact the brain. Deep Learning is supervised learning using extremely large neural nets.

This is the stage where biology is. What Charles was saying was that in order to get to the next step, someone has to create an algorithm to discover the specific gene to create the specific outcome, find the data, have a computer analyze that amount of data, and then have a laboratory to do the trials.

My next step was to find people who were on the cutting edge of that work. Few are doing this in their garage. They all worked for Google, Facebook, or other companies dealing with the high end of AI. While they didn't trade secrets, many were remarkably open about the work they were doing, not unlike Apple and Microsoft working together on projects in their early stages.

I checked my computer and found the International Conference on Artificial Intelligence was taking place in New York in three days. I quickly signed up. It was a three-day function with countless speakers. I started to look for the right people. There was a list of papers that had terms I'd never heard before. Over two hundred presentations were possible. I read through all of the titles. Only five were possibilities. I would certainly be the only person there that did not have a clue. It was clear from the categories that the world of AI had come a long way, and there were a lot of different areas. No one could be an expert in all of them.

I read the abstracts of the papers. One stood out: *Chaos Theory, Alchemy, and a Treasure Found: Break my Algorithm*. The presenter was Michael Longstreet. There wasn't a lot about him on the internet. He'd covered his tracks. There was mention that he'd been on the ground floor of Deep Mind, the AI division of Google. He also did work for DARPA, the Defense Advanced Research Projects Agency. I knew they did a lot of cutting-edge research – that was about it. I read the two papers being presented by people from DARPA and decided to sit in on those as well.

I knew something about Chaos Theory and a smidgeon about Alchemy. Longstreet was clearly a smart person, and the presentation looked interesting. I noticed it was being held in one of the larger rooms. They must know something I don't. The goal would be to ask him out to lunch and talk about what Charles had suggested. Perhaps he would have a few ideas. Like Clint, he stayed out of the limelight. I watched two of his TED talks and tried to read a few of his 85 papers. He did write a few op-eds or articles for the common folk.

Now and then, I would get in a research groove. Holding up in my apartment for days, reading, and writing for up to sixteen hours per day, I would go down the rabbit hole. Over the three days leading up to the conference, I printed, read, underlined, and took notes on over 800 pages of material relating to AI. Along with being a decent writer, I had two gifts with which I was blessed. The first was an almost photographic memory. The second was what General AI was about, connecting dots between disciplines. I did that well.

My first ten hours were taken getting the history of AI, beginning in 1956. As the Wide World of Sports used to say in their opening, "The thrill of victory and the agony of defeat." AI had known both. Bill Gates, in 2004, had said that when a breakthrough came in AI, that company would be worth ten Microsofts. While AGI has not been found, the leaps in machine learning have catapulted many businesses to astronomical growth, including Amazon.

The rest of my time was taken trying to understand the lingo and what questions I could ask Michael. That was if, a big if, I could get to him. I'm sure he is a very popular person. While no one knows his net worth, it is said to be close to a billion. He had three startup companies that each sold for over a billion. He is called on to consult with experts, CEO's, and board of directors. He has his pilot's license and is known to drive to an airport with people following him, get in his plane, and disappear. People say he meditates an hour a day, plays piano or guitar for another hour, works out for an hour, cooks, and likes going to small churches. On average, according to the two interviews I could find, he slept about six hours a night. He is also said to be an extremely generous human being.

8

I finished my work at 11 pm the night before the conference. I decided I'd had enough and would get a good night's rest. As fate would have it, Michael was talking on the first day. My guess was he could pick his spot. The conference was being held at the Marriot Marquis in Time Square. Lots of hotels, restaurants, and extra-curricular activities for the attendees. Close to 2,000 people had registered for the conference. I had no idea so many were interested and knowledgeable in AI.

At the registration table, the woman signing me in made a comment. "You are one of the lucky ones, Mr. Armstrong. Rumor has it that Michael will be announcing something special. When he does that, usually it is something remarkable."

"Then I am glad I signed up for it."

"Enjoy the conference and your stay in New York."

"I live here."

"You are the lucky one," she said, smiling.

"Yes, I am."

The energy in the hotel could probably run the air-conditioning. Every square inch seemed to be taken over by the conference. There were exhibits in the atrium and many of the 70 meeting rooms.

Within the AI community, there are segments. Probably via email or texting, they found each other in the chaos of the moment. I could overhear conversations about where to have dinner, theater tickets, multi-level learning, differentially private iterative gradients, business propositions, and new products that would be on display. While I heard multiple languages, the conference would be in English. There were a few workshops that would be in another language, and it was noted that in those situations, translations would be given in English. I wondered if I went to a conference in China if all the workshops and presentations would be in Chinese.

I knew absolutely no one at this conference. I only knew a few by name because of their fame or the papers I'd read they had written. Looking at all these people, I had a humorous idea for a computer program. What if a computer knew all the people I knew in the world. They also knew all the people those people knew. That went on for many iterations, let's say six — six degrees of separation. A computer should easily be able to connect the dots. My guess is at least five percent of the people at the conference were connected to me within three degrees of separation. We were not all that far apart.

The event started with a keynote speech by Demis Hassibis, one of the founders of DeepMind, the Google company focused on AI. He gave a spirited talk about AI, DeepMind, where things were headed, and the pace of research. He stated that while progress was steady and consistent, they were still looking for the breakthrough into the world of AGI. Demis pointed out that some people had said it would never happen. Others said it would be at least another 80 years, but the majority felt the first breakthrough would occur in the next 30. He commented that perhaps Michael Longstreet had found it. People laughed. He noted that fifteen presentations would be given about what DeepMind was working on. Everyone knew that papers were not presented by anyone on what was actually happening, but about what was already public knowledge. Guarding trade secrets and research was critical in the corporate and academic worlds. His talk ended, and there was a short break between sessions.

I wanted a good seat, so I went to the room where Michael was going to talk. I'd written a note to Michael when doing my research. I invited him to lunch or dinner at a location of his choosing, anywhere, any time. I let him know of my concern about the downside of AI and a little of the research I'd done. I bent the truth a little on this front. I got my card with contact information on it. I also told him to look me up on the internet. I probably didn't need to do that, but wanted to be upfront. My guess was I had a 30% chance of a response.

I walked into the room, and it was already half full. I walked to the dais and saw him shaking hands with people. I'd learned in my travels to be a bit bold, if not brief.

"Excuse me, Mr. Longstreet, I just wanted to give you this." I handed him the note.

He gave me that quizzical look. "Thanks."

I couldn't tell if it was a question or a statement.

Michael was an African American, about six feet, three inches tall, and quite thin. His hair was short, he dressed very casually but neat, and looked like the few pictures I'd seen on the internet. I decided to be optimistic and believe before long we'd be on a first-name basis. He was in great shape, and I could tell just by looking at him he suffered no fools. Michael had a winsome smile and quite a loud laugh. While he was the focal point of attention, he seemed to listen to those talking. I looked forward to getting to know him better.

The leader stepped to the microphone.

"Ladies and gentlemen, if you could find your seats, we will get started. Thank you for attending this year's International Conference on Artificial Intelligence. We have 81 countries represented, 18 universities and colleges making numerous presentations, 39 businesses sharing what they are up to, and of course, speakers like our next presenter who come representing themselves. Michael Longstreet is known to everyone in this room. He needs no introduction. Like E.F. Hutton, when he speaks, people listen. I will not take any more of his time. Please welcome Michael Longstreet."

There was a standing ovation as he took the stage and encouraged people to sit.

"Thank you for your kind words and greeting. E.F. Hutton lasted for almost a hundred years. They were last listened to in 1990. I can only hope my words and deeds make it a bit longer than that." The audience laughed. He was a legend, and no one doubted he would be a person of history.

"My remarks today will be brief. But I do come with a challenge that will earn someone some money, and hopefully will move part of the AI world in a different direction.

"Our world is fraught with issues. We seem to resolve some of them while others linger, and new ones arise. The ongoing struggle of the have-nots continues as we in the have department do better and better. Don't worry, I'm not here to talk politics. I do believe AI can make a difference on virtually all levels. We are constantly pushing the boundaries of Machine and Deep Learning."

"Over the past two years, I have been delving into the worlds of Chaos Theory and Alchemy. You might ask why. I looked at the mathematical notion of the Butterfly Effect, Bifurcation, Strange Attractors, Topological Mixing, Self-Organization, Fractals, and Complexity. I have also been researching biomimicry or how we can learn from nature, the foundation of all systems. Alchemy has long been an interest of mine for several reasons. The first is that alchemy has been around for thousands of years in a variety of forms; it just doesn't go away. Most people get hung up on the idea that alchemists were trying to create gold out of anything: the Midas Touch. While virtually all alchemists had some type of laboratory, they were more interested in the psychological and spiritual dimensions of the process. They looked at how the natural world could be mimicked in our internal processes. Heating things up, distillation, separation, coagulation, and several other processes apply to the maturation of all systems."

"I have come to believe that while we make gains in AI, we are mostly stuck on a linear path. Chaos Theory, Biomimicry, and Alchemy, all show methodologies to get out of the box. We aren't certain who coined the phrase, but we do know a few things. In 1914, Sam Loyd's *Cyclopedia of 5000 Puzzles, Tricks, and Conundrums (with answers)* contained the "Nine Dots Puzzle. It was simple. Draw a continuous line through the center of all the dots so as to mark them off in the fewest number of strokes. He forgot to add that the lines had to be straight. That does make a difference. Here it is. Many of you have probably seen this."

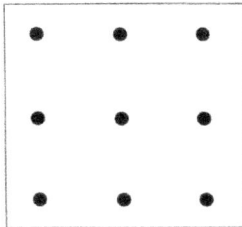

"Most of us go crazy thinking about this, and we go through many iterations. It's only when you think outside the box that the answer arrives."

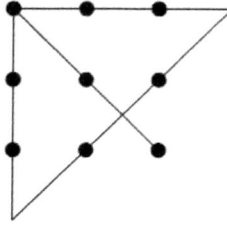

"There are many ways of doing this, but you must go outside the box to do so. The rules didn't say you couldn't broaden the envelope."

"Edward de Bono coined the phrase Lateral Thinking in 1967, another way of saying the same thing."

"John Adair, a British academic, claims to have first used the term in 1969. This debate goes on. Who cares?" The audience laughed.

"I want to encourage all of you to experiment with wild ideas. Niels Bohr is quoted by Freeman Dyson as saying, 'We are all agreed that your theory is crazy. The question that divides us is whether it is crazy enough to have a chance of being correct.' He said this to Wolfgang Pauli after he gave a presentation of Heisenberg's and his own nonlinear field theory of elementary particles at Columbia University in 1958."

"Yes, we will continue to move forward. Yes, in the scope of human history, we are moving at dizzying speeds. The number of problems we want to solve grows by the day. Our brains are remarkable, but they are slow. What lies ahead has almost limitless potential."

"Progress always comes with a warning. As we created programs for computers, the darker side of humanity joined the race and created malware. Some of us find ways of stopping the latest malware attempts while others spend their time trying to thwart the latest security. We all know that we are not safe."

"The faster and deeper the computing power, the closer we get to AGI. The danger is that when AGI arrives, the shadow side of our nature will also show up. I do not worry, at present, that computers will take over the planet and dominate us. I do have grave concerns about using AI in warfare, biology, in our economic structures, and in virtually all aspects of life."

"I have always found it humorous that we want a computer to think like a human being. The Turing Test still stands, as does Steve Wozniak coffee test. Unfortunately, humans make mistakes. Yes, our brains err. My guess is that every one of you has made multiple mistakes in your work. Mistakes are part of learning. Unfortunately, in certain areas of life, one mistake can be disastrous. An autonomous vehicle or weapon can kill or maim a great many humans. Yes, humans kill other humans, at times by accident, and at other times on purpose, as in war. I am not against autonomous machines. I simply believe we need to be careful where we put responsibility."

"Let's wrap this up. I am not saying anything you haven't heard before. I do offer you this challenge. The website you see on the screen is an isolated computer. It is the fastest computer you can purchase online. The software is special, not the computer. You are no going up against a large machine. You say it isn't isolated because it is on the net. The first challenge is for you to see if you can access this computer - not the website, the computer. We will use the capture the flag technique. You will see the message I want sent and where I'd like it put. The second challenge is to see if you can put a virus into the machine of any kind. I have the program and can easily see if anything has been added or deleted. If I find one change, I will know someone was successful, and I will write them a check for $5 million dollars. I see some of you already opening your computers. I strongly suggest you wait a couple of minutes."

"How can I do this? I have developed an algorithm based on the premises I have discussed. Actually, a series of layered algorithms that mimic viruses, algae, and the human brain. My program thinks outside the box, on its own. I believe this is the next step in AI. As with all things, it needs to be tested. That is where you come in. If I am successful, I will publish my strategy within the year."

"One final note. This is a program for computer security. The idea is to protect computers and networks from malware. One of the ways this system works, is in its ability to trace where the malware came from. You can try to reroute what you are doing, but it understands VPN software. When it does so, my computer takes the malware program, rewrites it, and returns it to the sender. It also makes note of the date, time, and source of the attack. How does it do that? It thinks - very fast, and is not trapped by linear oriented code. This pushes Machine and Deep Learning several steps further and is on the edge of AGI."

"I wish you well, but realize that you will be signing an affidavit when you log onto my computer, releasing me from all liability when your computer is attacked by a version of the malware you sent. I encourage you to use a computer that is not attached to any network, and that has nothing on it other than the program you load. My computer will do nothing to your computer if you are poking around. When you attempt to upload to or download from my computer, things will happen. You might say that this is not fair, that trying to break into a computer from a simple laptop is barely possible, especially on this scale. I can't disagree. If you choose to use Watson, Summit, or other mega-machines, be my guest. When you bring malware to a billion-dollar machine, that will be on you. Before I release the software, I will let the big machines have an opportunity to try with no repercussions."

"I appreciate your listening, and I look forward to hearing some amazing presentations over the next few days."

At first, I'd thought Michael was a 5/6 on the Enneagram, highly rational and very loyal. Now I believed he was another 7/6, good at a lot of things, bored easily, and loyal to the causes he believed in. I felt he was pretty healthy, which was good news. Unhealthy 7s were not fun. I looked forward to seeing the talents he had.

Michael received a standing ovation. Apparently, he never took questions. He walked off the dais. I watched him shake a few hands, and then my mouth dropped. There, standing by the exit door he would walk through, were Clint and Mary. What were they doing here? I am rarely stunned, but this did.

9

Michael looked at them with a puzzled look. They started chatting, then he walked through the door, and they followed. I needed to hear what they were talking about, but that was not going to happen. I started to follow them. Then I stopped. At first, I took a few seconds to think about what I'd say. There would be no reason for me to be following them. Mary would wonder what I was doing there. As the pondering went on, I noticed something else.

In my line of work, you learn to be attentive and paranoid. I have become pretty good at picking those tailing others out of a crowd. This one was almost too obvious. Tall, well built, in a suit which was one of the only ones in the hall. He was talking to his microphone as he headed through the door. I quickly looked at other exits and saw at least one other tail. I suppose it could have been a coincidence. That thought dropped quickly from my mind. Were they tailing Michael, Clint, or Mary? Clint already had a tail in Boston. I'm not quite sure how they found him here. Michael was easy to find; he was speaking. I knew of no reason to tail Mary. Now all three were gone. Then I remembered.

I pulled out my computer and sat at a table. I had not been tracking Clint since I left him at the farm. The map popped up, and the two blue lights were blinking. One was attached to his car, not far away, and one was in his satchel. I gave them five minutes and followed. They went into one of my favorite Italian restaurants, Buca di Beppos on 45th just off Time Square. I was thankful it was a busy place, so I could go unnoticed. I saw where they were seated and chose a table where I could use another of my toys: a directionally sensitive microphone. Looking around the restaurant, the tails were spotted as well.

I set my microphone out, plugged my earbuds in, and ate lunch as I listened. My table positioning was such that I blocked the straight view of their table from the feds, but they would still probably hear enough.

"I will say that what you said at the conference got my attention, so please say more," said Michael.

"Mr. Longstreet."

"Michael, please."

"Michael, you are certainly one of the leaders of the AI movement, and what you said today at the conference, and in some of the papers you have written, shows you are concerned about its future," Clint said.

"You've read my papers?"

"All of them. Do you know much about me?"

"I have seen you on TV a couple of times, but have read no papers or seen any full interviews. To be honest, genetic engineering is not an interest of mine."

"Perhaps I can pique your interest. You are concerned that AI, when advanced, may either do things we don't want computers to do, or someone can use AI for darker strategies. In the worst-case scenario, Terminator."

"Bluntly, yes. Obviously, it is more complicated than that."

"I understand. And you believe we are perhaps twenty to thirty years from that moment, if not sooner."

"Yes."

"The trajectories between our two fields of interest are on a direct collision course. You think that scientists around the world are simply trying to find vaccines and cures to smallpox, malaria, AIDS, Covid-19, and whatever comes next — because there will be a next. You must know..." Damn, I thought. The waiter was standing between them and me. I was missing important pieces of information.

"Yes, and I don't think I am alone in that."

"So there are nefarious computer programmers creating bad malware, but all bioengineers are benevolent, right?" said Clint.

"Now that you put it that way, probably not. But gene-editing is a long way off."

"Before the experts in the field walked into that auditorium today, they thought what you presented as a reality was far off. Let me tell you the reality. There are……and those are the ones I am aware of. Think of what……. in five years." I continued to only get parts of the conversation.

"True."

"By the way, this is Mary, my assistant."

"We met briefly," replied Michael. I could tell he was wondering why she was here. I also decided I'd better start calling him Michael.

"Let's just say that things are not as far off as one might think. I have an ….. things down on both fronts, if not….."

"I would love to hear that. Is it ….?"

"We think the first step would be…" I took out my earbuds and blocked the view of the feds.

At that moment, I decided the less they said, the better. If the feds heard them plotting, their lives would be in danger. I'd had the feeling this might happen, so I'd penned a note while waiting for my tea and lunch. Fortunately, theirs was arriving, so they stopped the serious conversation.

I got up and walked over to their table. They all looked at me.

"Thanks again for the wonderful speech. I wish you well in all you do." My back was to the feds, and I put the paper on the table in front of Mary, since she knew me. She tried to act nonchalant, but I could tell she knew something was up.

"It was my pleasure, thanks for coming. I will give you a call," said Michael.

"I appreciate it," I replied, shocked he'd remember me. Not necessarily a good thing in light of what was happening.

I went back and sat down. Mary started to read the note which said: "Please believe me when I tell you this. Don't look now, but the two men sitting at the table behind me are federal agents. They are listening to your conversation. They have been tracking Dr. Williams since he went to Boston. Switch the conversation to something else. When you get to Dr. Williams' car, check behind the license plates, that is often where they put tracking devices on those who don't expect one to be on their car. Move the conversation to things in your fields, but not what you actually are here to discuss. If you find the tracking device, put it in the same place on the car next to you. DO NOT LEAVE THE TABLE until you are done with lunch. They will know something is up. DO NOT look at them."

Mary glanced quickly at me, then back to her lunch and the others.

"Michael, what part of the world are you from?" asked Mary.

He gave him that quizzical look that said, "Where did that come from?"

"I have a few places I work, and I have a small place here in the city."

"Are you a Yankees fan?" she stated as he slid the note to Clint, who was going to flaunt it till Mary put his hand on it. Clint got the message.

"Um, no, Mets." He saw Clint reading the note with some surprise on his face he quickly wiped off.

"What do you do for fun when you aren't working with AI?"

"I like to fly fish and travel. I am also a woodworker."

"Mary loves to travel, as well. Tell him about your latest venture," said Clint. There was a tablecloth that fell well over the sides of the table. Clint passed the note to Michael, feeling like he was back in middle school. Michael took the note with a little puzzled look on his face. He'd watched both Mary and Clint read it, so he knew to be a bit careful.

"I just returned from China. I went to a lab in the middle of nowhere. They are very curious about how these viruses are being created in the wild and transferred to humans."

"Are they discovering anything?" Michael didn't want the subject to totally change. He took out a pen and scribbled a note on the paper, and held it.

"They haven't figured anything out yet, but they are looking into protein folding."

"You lost me." They laughed.

"Sorry," said Mary.

"No worries, and don't try to get me caught up, I will just trust you. Sounds fascinating. Do you remember this diagram in my talk?" He passed them the note.

The note asked if they believed what the other note said. He glanced over at me and the feds. I couldn't blame him, and it was human nature to check things out.

Mary and Clint nodded. "Yes, we do."

"So, there are protocols and conventions about biological warfare," stated Michael.

"And about interfering with other countries' businesses and governments, I believe," responded Clint.

"Touche."

"Somehow, we have to help our government or the U.N. understand we need to do something," said Mary.

"I don't know how I can help. I'm not a lawyer or a politician. I write programs. I have just created a program that may change everything."

"In the world of viruses, once you have the virus, you can create the vaccine, it takes time, but it can be done."

"Have you found a vaccine for malaria or AIDS yet?" asked Michael.

"No, but we believe they will be found."

"I know enough about them to know they are both very clever at avoiding the immune systems and vaccines."

"That is very true, but we have some new ideas. If someone gets ahold of your program, can't they find a way around it?" Mary inquired.

"Yes, but they are stopped before they enter the front door. That is the magic behind this program. I have gone not only out of the box we live in but off the paper."

"How big is the program?"

"Currently, about 25,000,000 lines of code."

"That is a lot of code. How did you have time to do that? I think the old Windows 10 was 50 million lines," asked Mary.

"That is the beauty. I didn't. The computer did; I just gave it the ability to do so."

The conversation went on for another twenty minutes. Clint passed the note back with his phone number. He too was paranoid and smart enough to know he was watched on some level, and his phones were probably tapped now and then. It comes with the territory in the area of work he does. He gave him a piece of paper with his regular number.

"Call me. We do appreciate your time. Give it some thought, and let us know what you think," said Clint. They shook hands after paying the bill and leaving a hefty tip and left.

At the front door, Michael told them he needed to make a call, and he hoped to see them at the conference. They were on their way out of town, only having come to the conference to see him.

"We are doing a little shopping, then heading back for another presentation."

"It was a pleasure meeting you, take care," said Mary.

"You too."

The feds gave them a little head start and then followed. I could see that Michael breathed a sigh of relief. I'd not expected what had just occurred. You'd think I would learn to expect the unexpected and be prepared for everything. That would take a moving van. I was wishing I had another phone to give him, but I didn't.

Walking out the door, he stopped me.

"Who are you?"

"Tom Armstrong."

"What is going on?"

"Now is not the time. You have my number. Will you be at the conference later today?"

"Yes, I will be heading back in two hours to hear a talk a friend is giving."

"I will meet you at the door with a phone. Only call me from that phone."

"What is going on?"

"You have a computer, look me up. There may be nothing going on, but those feds think there is," I said.

"Now, you have me a bit worried."

"Dr. Longstreet."

"Michael."

"Michael, my guess is the feds have had close tabs on you for some time. There are probably spies in the organizations you work for, your phones may be tapped, and your computers as well."

"They are protected."

"I hope so. Dr. Williams is a good man, but he may be treading into something he does not comprehend, just be careful."

"OK."

I walked away, knowing he was more fearful now than before we'd chatted. I also knew I'd be seeing him again soon. Now I needed to buy another phone.

10

First, I went back to Clint's car. I was hoping they went to the conference and not the car. They wouldn't find a tracking device behind the license plate because I'd already removed it. The feds were either tracking him twenty-four hours a day or had tapped his phone. Probably the latter. I decided to take the risk and went into one of the many electronics shops near Times Square.

I always stood in awe in the Square. The advertising signs seem to grow in number and size by the day. Now and then, a picture of someone would show up with a happy birthday, or will you marry me, caption. You knew there were somewhere in the Square. I couldn't get that distracted. I bought two phones, just in case something else came up.

I jogged to the garage where Clint's car was parked. My computer did well at tracking on the surface. I wasn't sure how accurate it would be going down a four-story parking lot. I started walking. The worst-case scenario was that I would walk down five flights of the garage. I would have to pass it at some point. The computer had a blue beeping light, so I knew the car was here. On the third floor, I found it. My device was safe, and I didn't think they'd look for it with something else to go after. I took a cheap tracker out of my bag and attached it to the back of the license plate. That job was finished.

When I entered the lobby of the hotel, there was an enormous crowd looking at a giant monitor. Somehow, someone had put together a program showing the activity on Michael's website that people were trying to hack.

On the left of the screen were the rules.

1. Enter at own risk, if you enter, I have no liability as to what happens on your computer or the network to which that computer is attached.
2. Poke around all you want. If you attempt to download or upload a file, something will happen to your computer within the next 24 hours. The more potentially dangerous your actions, the more damage will be awarded to your vehicle of destruction via the malware or virus you sent.
3. For the successful person, I will write a check for $5 million.
4. I will tell you that there is an answer. When or if I reveal the methodology, you will know.
5. I have written a program so those interested can track what is taking place on the website. Those statistics are to the right. The program can be found at zipattractor53465gb.net.

Now I knew where the program came from. The top box listed the number of visits to the website. In the past five hours, there had been 927. Below that was a list of attempted and successful attacks. There had been 113 attempts and zero successes. Under that statistic were the grim details of what happened to those 113 attempts. I had no idea a computer could do this. The graph on this screen only listed the most recent ten attempts. On an individual computer, you could scroll down and see what had happened to all of them. The top five I saw were to disable the mouse and keyboard, delete emails on the computer and in the cloud, have the computer programmed to blink off and on every two seconds, and change the order of a qwerty keyboard so they didn't know what key actually produced what letter – and that would change every two days. Finally, one computer could not stop working at its hardest, overheated, and shut down.

Those watching stood in awe. There was more interest in how Michael had gotten this far than how to beat it. I tried to listen in on conversations. Most people either felt he had access to immense amounts of computing power, like two IBM Watsons, or he'd cracked the foundation of the AGI algorithms. Some believed he'd created the first biological computer. Of course, there were the conspiracy people who felt he'd joined forces with the government, and soon, all of their computers would be under the thumb of Big Brother. Most people laughed at that, knowing the typical person's computer was not that difficult to hack. Everyone knew there were backdoors to Windows and Apple. Very few people knew how to find them. I had friends at both companies, and they acknowledged they were bombarded by attacks – from individuals, other companies, other countries, and our own government. The information held in the databases of those companies plus Facebook and a few others boggled the mind.

I checked my watch, and it was time to meet Michael again. I'd changed my shirt and hat to hopefully not attract the attention of the feds if they were following him. This was a smaller crowd, probably one hundred people attending the talk, which was on *International Cooperation Around Cyber Warfare*. Three minutes before the start of the session, Michael walked to the door. He looked around, checked his watch, and probably decided I was a nutcase and went in. I followed quickly, bumped into him and put the phone in his coat pocket, and kept walking. He knew what had happened and showed no signs of awareness – a good thing in this situation.

I went and found a quiet corner of a nearby café. I'm not a coffee drinker other than the one with which I start my day. I ordered an iced tea. Deciding a few weeks ago that I put away far too much diet cola, I stopped cold turkey. Caffeine-free was all that found its way into my body now. In the heat of the summer, a good iced tea sat well. Later in the afternoon, that would become a long island iced tea. The best.

I pulled out my "Mary" phone. There was a message: "What were you doing at the restaurant, and what is happening?"

I responded, "My job. Call me. This is getting serious way too fast."

"Give me 25 minutes," she texted back.

I took the time to attempt to get a bit smarter. I knew that DNA was a double-stranded helix-shaped molecule that determines how we are created and function. DNA and RNA work together to create proteins that underlie all functions of the body. They are both polymers made up of sugars, phosphates, and bases. RNA has Uracil rather than Thymine. Uracil takes less energy to create.

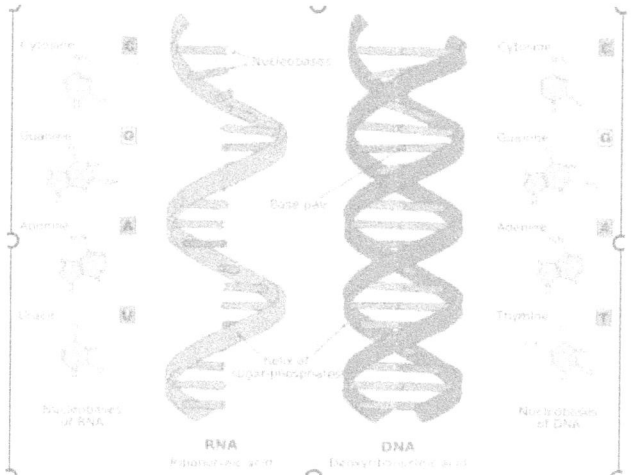

Wikipedia: Nucleic Acids

I understood enough to know that if they were going to solve the virus issue, they'd need to work with DNA and RNA. I didn't quite understand the difference, so of course, I looked it up. I thought back to my college days where 1) some of this was still unknown and, 2) we didn't have computers, so it would have taken me hours or days to track down this information. Today I pushed a button, and 200,001 results came up on one search engine.

DNA replicates and stores information about the genes. It's like a blueprint for the organism. RNA converts the information in DNA into a format to create proteins, and then moves those instructions to protein factories in the ribosomes.

Each DNA molecule has two strands that are made of nucleotides. Each nucleotide has a phosphate, a nitrogenous base, and a five-carbon sugar molecule. For perspective, human DNA has roughly three billion nucleotides. Bacteria have around five million. The Covid-19 virus has 29,903. Humans are very complex creatures.

In the nucleus of all cells, we find chromosomes. They are strands of DNA wrapped within proteins called histones, in part to keep them from getting too jumbled. Chromosomes are divided into genes, segments of the DNA that have a specific designated function in the organism. It is the conversation between the DNA and RNA that determine when and how those genes are manifest. It is also the place where many mutations occur.

Molecules are paired together to create base pairs. A large RNA molecule may be a few thousand base pairs long. They do come much shorter. Most genes are around 27,000 base pairs and can be up to two million. In DNA, the base pairs are A-T (Adenine and Thymine) and G-C (Guanine and Cytosine). In RNA they are A-U (Uracil) and G-C.

Both of them are found within the nucleus of a cell. However, the RNA is found in a distinct part of the nucleus called the nucleolus. DNA roams around outside the nucleolus.

There are three types of RNA. Each has a specific function. mRNA copies parts of the genes, transcription, and moves them to the ribosomes, the factories that produce the proteins. tRNA gathers the amino acids and brings them to the factory. In essence, they translate the recipe the mRNA brings and gather the ingredients. rRNA takes the ingredients and makes it happen.

Mary's phone rang.

"Hi Mary, how are you doing?"

"How do you think? The government is spying on us, an investigative reporter is spying on us, and my boss seems to be going off the edge." I could hear the anxiety in her voice.

"I understand but take a deep breath. You need to understand you can still walk away. I don't think you've done anything illegal, and I'm sure Dr. Williams would understand." There was a long pause. "You haven't crossed the line, have you?"

"I'm not sure, I know we are close."

"What does he want with Dr. Longstreet?"

"He knows Longstreet is concerned about the spread of computer and biological viruses into the wrong hands. Clint thinks Michael can help streamline and fast track the process," said Mary.

"To what end?"

"He hasn't shared that with me. He has set up a separate computer system in the lab that is not connected to anything. He is also disappearing for periods, and I have no idea where he goes."

"I am going to give you an address of someone to go see to give you some tools that will help you feel safer," I said.

"I don't need guns."

"They aren't guns, relax. Do you want to know if you are being followed, having your phones tapped, or other wonderful things your government and others do?"

"I guess."

"Go see Jim. He's at 52a West 48th. Tell him I sent you. He will have things ready."

"OK," she said a bit nervously.

"When will you meet with Michael again?"

"I don't know if Michael wants to or how we do it without being followed."

"I am confident he wants to. Leave the how to me. He is about to get out of a talk in room 154. Go ask him. If he wants to meet with you and Clint, text me. You have to go now if you want to catch him. By the way, I did not know you were going to be at this conference, I came to see Michael. For the same reasons you apparently did."

"Really? I am in the lobby," Mary said.

"Yes, OK, and let me know."

I was getting this sinking feeling that these two geniuses had been working on things for some time. I had this intuition that if and when they decided to collaborate, the end result would be stunning, frightening, and arrive on the scene quickly. It was difficult for me to imagine what it could be. I didn't see either of them as doomsday people that wanted to destroy the planet. On the other hand, I knew they were starting to get desperate.

Here is a rhetorical question for the reader. Let's say you knew about a company that was bad. Everybody knew it was bad, but no one did anything. Then you found out someone who you knew was going to threaten to blow up their main plant unless they stopped producing the horrific chemicals. You believed it to be just a threat, but you also knew the person was on the edge and might actually do it. What would you do? You believed in the cause and in the person but weren't sure those were the right methods – even though everything else had been tried. I was not in that situation yet but felt myself starting to slide that way.

Thirty minutes later, the phone rang.

"Hi Mary."

"He wants to talk."

"I thought so. Here is what you do. You and Clint go to your rooms and change clothes. Then you leave. Carry nothing with you other than cash for the subway and something to eat, and I mean nothing. They can put trackers on anything. Go out one of the side doors. Take the subway to 110th and walk to the Cathedral of St. John the Divine. Go into the Peace Fountain area. There are plenty of places to sit and chat. You won't be followed or listened to. Let me know how it goes. I give you my word I will not be recording anything nor will I be anywhere in the vicinity."

"Thanks, I will."

11

An hour later, they were all at the Peace Fountain. The fountain sat next to the cathedral, one of the largest gothic structures in the world. Depending on how you define the size of a building, it is the largest cathedral in the world. The grounds of the cathedral are peaceful, well landscaped, and an oasis of spirituality.

"Hello Michael, thanks for coming," said Clint.

"I could hardly say no with all that has happened."

"How are people doing trying to get into your site?" Mary asked.

"As I expected, not well," he said with a smile on his face. Before we start, who is this Tom Armstrong person?" asked Michael.

"Who is Tom Armstrong?" inquired Clint.

"The man at the restaurant," said Michael.

"No, his name is Tom Lassiter. I met him at a Senate hearing," Clint stated.

Mary bit her tongue.

"How do you know his name is Tom Armstrong?" asked Mary.

"He handed me his card and said he wanted to have lunch. That was before the restaurant. What is the connection? Why is he following us?" asked Michael.

"Let me check." Mary pulled out the phone Tom had given her.

"We were told not to bring phones," said Clint.

"You can relax. This is a phone I bought to make calls I don't want the government to know about. Working at Fort Detrick, I know they check phone records to see who we are contacting. Finding calls to foreign embassies would not be a good idea," Mary said.

They stared at her.

"Yes, I'm paranoid, sorry." She typed in the name Tom Armstrong and came up with hundreds of hits. "OK, he is an investigative reporter. He is well known for putting political and business figures behind bars. I see articles on a wide variety of subjects. Lately, he's been investigating election issues. There are six books in print, all have sold millions of copies. He doesn't seem to like the limelight. I can't find much about his personal life. He lives in New York and is divorced."

"What does he want with us?" asked Clint.

"I suggest the next time one of you sees him or talks to him, you ask," said Mary.

"We do owe him one for the restaurant," said Clint.

"Unless, of course, it was a setup – to build trust," Michael replied.

"Paranoia builds fast," Clint stated. "Anyway, I wanted to talk with you for a reason."

"And what is that?"

"As I stated at the restaurant, we have similar and converging issues we are concerned about. I will talk about the biological side of things. Then I would like to hear about the AI side. I am sure you know that Fort Detrick is the center of research on the defense side of biological warfare. We look at how to defend against attacks using the bacteria and viruses known to humanity. Fairly recently, I was notified that other countries were putting resources into developing new bacteria and viruses. Think about it, what if you had a virus and the vaccine that the world had never seen. Suppose its rate of infection was astronomical, and the mortality rate over 75%. For perspective, Covid-19 was less than 3%. You know that certain viruses, like HIV, have no vaccine. They mutate and avoid the immune system. It took twenty years to come up with medications that hold it at bay."

"My superior ordered us to develop one of these viruses. None of my colleagues from laboratories in other countries will acknowledge they are doing this, but I have no doubt they are. It seems we have entered a new cold war. MAD now exists on a new level. With the atomic bomb, we could set it off on an island for the world to see. This is more difficult with a virus, but possible, especially if you are ruthless."

"How is that even possible?"

"Like in your world, computing speed and power have changed everything. As you might expect, working with Fort Detrick and with DARPA I have access to the best of both worlds. I have a black budget of hundreds of millions of dollars. We are currently experimenting with malaria, one of many tricky parasites."

"I know nothing about genetics. Why did you contact me?"

"Computer viruses, as I understand them, are not that different from biological ones. They look for specific things in the host. Most believe viruses are not living outside the host. When they find the right cells to invade, they take over and spread rapidly. That would be like your viruses looking for a way into the main program or programs of the computer. Once in, they start to do the damage as well as to spread to other computers. Your computer coding is not unlike genetic coding. In any given set of instructions, there are weaknesses and places a virus can attach and invade. In genetics, we isolate segments of the billions of base pairs we have. We compare and contrast these and find out what their function is, what proteins they create. The problem is this takes time," Clint said.

"Let me guess, you want me to create an algorithm that will speed that process up, tell you where the weaknesses are, and what you can do to thwart them," Michael said.

"Yes, as well as to strengthen those weaknesses. You have found a way to look at everything as it attempts to invade the host, the computer. I would also like an algorithm that can search all databases and find specific information about two specific issues I will discuss in a minute."

"Yes, my program looks at the entrance of a virus, as do most antivirus programs. But my program also looks at replication and transmission protocols, which viruses need to do their damage and spread the news to other networks and computers. I suspect, like your viruses, ours are not one line of coding. If you know what to look for, they stand out."

"That is miraculous."

"Close to it."

"That must be an enormous program," replied Mary.

"At this stage it is, but I have set up a separate neural net to allow the computer to find ways of shortening it. While Einstein's $E=MC^2$ looks and sounds great, there is a massive amount of mathematics behind that."

" Got it. Viruses in humans and animals enter cells and change how proteins are created. Proteins are folded in very complex ways. Think of it like origami. You have a piece of paper, nothing special. I have a hard time making a crane, one of the more simple forms. An expert can make anything, including a musical organ. It's all in the folding. What if we work backward? We take the end result, unfold it, find the sequence of amino acids that make the protein, and then the DNA and RNA segments that create that," said Clint.

"An interesting idea, I will give you that."

"Why viruses rather than bacteria or other parasites? You'd mentioned malaria."

"Excellent question. Mary, you tell him."

"OK. Viruses have between a few and 200 genes. Bacteria have between 500 and 1200. The parasite that causes malaria has 4,300. Less is more. Isolating and changing the DNA in the malaria parasite is difficult. Over hundreds of millions of years, it has learned to adapt and avoid detection. Proteins help the parasite hide within the mosquito and within the human. Understand that the mosquito is responsible for more death on the planet throughout history than any other entity, including humans. They have recently discovered the protein in the mosquito that protects the parasite, *Plasmodium falciparum.* Researchers have been trying to find ways of changing how that protein folds to unprotect it. The answer lies in finding the gene that makes that protein. We have been experimenting and have narrowed the field down. We believe we are close to finding a way to have humans create a similar protein that, like the mosquito, would protect us. This mechanism may be true for many diseases we currently have, as well as ones that may arise in the future."

"I am still a bit unclear how AI could help."

"I said I had two things I wanted you to work on with us. Here is the first. We will soon have the machinery to do thousands of tests a day and isolate gene segments. What we don't have is the ability to translate that data into a form computers understand and can compare to other genes and mosquitos to track down the right segment. Once we find that, we can change the segment and make it incapable of producing the protective protein."

"But you can't do that to every mosquito," said Michael.

"Very true. Only females bite and spread malaria. If we change the genetic structure of one, all of their offspring will carry that trait. Each mosquito will have between 500 and 5000 offspring in their short lives. We already know where the sex gene is, so we can produce virtually all females. We mass produce these engineered mosquitos by the billions and let them loose. In a relative short period of time, they will be harmless."

"There must be several species of mosquitos."

"Over 3,000, but only one carries the malaria parasite. Of course, we would also be working on the human side. If we can unlock the mechanism that keeps our immune system from seeing the invader, the parasite would not get far."

"You actually believe this could work?"

"Yes."

"OK, I'm in on that one. Now about the other one, concern over biological and cyber warfare."

"I'm a believer in not sharing information that will potentially cause you problems. How about we work on this for now, and as things move along, I will fill you in. I'm not convinced what I am thinking about will work," said Clint.

"Is it dangerous?"

"In our field, everything is dangerous; that is why we have four levels of biological security. One final thing," said Clint.

"I love it when they save the big one till the end," Michael said, smiling.

"I don't know if this is possible, but having you crack into our data center and create a way for me to access the data somewhere it can't be traced would be very helpful."

Michael took out a piece of paper and a pen and wrote some things down. He handed it to Clint.

"This is what I'd need."

"How do I get this to you?"

"Give it to me tomorrow at the conference."

"We are leaving today. The only reason we came was to see you."

He took back the piece of paper and wrote something else.

"Send it to this email, but do not use a computer that has any connection to you or anyone you know. Use a library or something. Do not attach your name. Create a new Gmail or Hotmail account with completely fake information. Do either of you have children?" asked Michael.

"I have a 7-year-old daughter. She is the love of my life, along with my husband," said Mary.

"My wife, Luly, is an artist. I have two children, both grown. My forty-one-year-old son has some special needs and lives in a group home. I see him a couple of times a week. My thirty-seven-year-old lives in Seattle. She works for Microsoft," said Clint.

"Why not ask her to do this work?"

"She is smart, but not in the same league as you, and AI is not her forte."

"Grandkids?"

"Nope," said Mary. They all laughed. "Clint, grandkids?"

"One grandson who is four. I try to see them as often as I can. You?"

"I have three children, two sons, and a daughter. Ben is 39, Joan is 42, and John is 46. Two grandkids. None of them are in the family business. One is a vet, one an actor who seems to stay busy, and one works for the CIA, doing what I don't know. Fortunately, in my line of work, I get to spend time with them often. I have been divorced for almost 15 years. No one can put up with me. I don't blame them," Michael said.

"I am constantly amazed that my husband puts up with me," said Mary.

"Michael, I am trusting you understand that if the world doesn't change pretty soon, bad things will happen, very bad," said Clint, having had enough with the small talk.

"Yes, I understand."

"Good, we are, at least for now, on the same page. Have you thought of a way to stop the craziness?"

"I have had a few ideas, but none of them good, and I hadn't given much thought to your form of disaster," stated Michael.

"Fortunately, at least for now, on your end of things, you can unplug. Am I right?"

"Yes, computers aren't at the point they control everything, but when we give them the ability to think for themselves, that will change readily. But you do remember the election of 2016 and Russian interference. 2020 was worse because they'd done little. You may have noticed that most states have gone back to paper ballots. But those are tallied and put into a computer. The banking system, the Pentagon, and various computer data centers are attacked daily. Billions of dollars are being spent to protect them. The hackers will eventually figure it out."

"In our world, once the genie is out, it's out. Viruses don't think; they simply engage, multiply, and destroy. Unless, of course, you have created a way to stop them. Smallpox is never going to be extinct. Fortunately, we have the answer. We don't have an answer to AIDS, Corona and other viruses, bacteria, and parasites," stated Clint.

Mary started to laugh.

"What do you find so funny in the midst of a discussion that could mean the end of the world?" asked Michael, a bit annoyed.

"I find Tom Armstrong an interesting person. I believe he sent us to this spot for a reason."

"And what might that be, other than it's outside and isolated?" asked Clint.

"Well, have you looked at what sits in front of us?"

Both of them looked at the bronze sculpture that stood before them. It stood forty feet and had numerous animals and creatures on it.

"What do you see?" asked Mary. "What is at the top?"

"If I didn't know any better, I would say that is Michael the Archangel defeating Satan who looks decapitated."

"Yes, the battle of good and evil. What else?"

Now they were on their feet and walking around the fountain.

"A sun and moon."

"Meaning what?"

"This might have been meant for me. The sun and moon are significant symbols in alchemy. The sun is the cosmic power of life. The moon represents death, birth, and resurrection," said Michael.

"Clint, what does the pedestal look like?"

He gazed at it for a minute and laughed. "DNA."

"There is something here for all of us. The small figures on the bronze rope around the base were created by kids, the hope and future of the world."

By now, they were all in their own world, interpreting the sculpture in ways that fit with their beliefs and understandings of the world. The silence was broken when Michael started to laugh. He was standing by the largest of the smaller sculptures that were wrapped in the rope surrounding the fountain. Clint and Mary walked over and started to read the inscription on the bronze plaque.

"That is not what is funny; this is," he said, holding up a piece of paper. "It says, 'Welcome my friends to one of the gems of the city, the cathedral, and this fountain. You might want to check out the cathedral if you have time. I hope your discussion was good or will be good, depending on when you find this note. I have no idea what you are up to. I can only hope it is good. You know how to find me if I can help. PLEASE, be careful. There is already increased security on all of you since you have met. Mary can explain further. The more I dig into AI and gene-editing and discuss it with my contacts around the world, the more nervous I become. Take care."

"I'm glad we found this first."

"I'm not sure what anyone else would have thought it meant," said Mary.

"What did he mean by 'ask Mary'?" Clint inquired.

"Back at the hotel, he found me and gave me a note with a name and address on it. I believe this person will have materials ready for us so we can see for ourselves if our phones are tapped, or our cars are being tracked." He wrote down a copy for Michael.

"OMG," said Michael.

"I think he believes this is serious business."

"I have no idea how he knows, but we all know he is right. There is no half-way on this. Each of us is either all in or all out. Michael, the Archangel, was all in. You ready to slay the evil one, Michael?" asked Clint.

"Unlike you, I don't see the evil lurking, ready to pounce yet."

"How long will you wait?"

"Good point. I am hoping to find defensive measures."

"And tell me how has that worked in the past at slowing humanity down?"

"Never."

"I agree," said Clint. "Currently, we are ahead of the game in each of our fields – as far as we know. If we get behind, playing catch up and having governments and companies stop will not happen. Do you agree?"

"Sadly, yes."

"Then, do we have a choice?"

"Not really."

"How much do we tell Tom?" asked Michael.

"I do trust him, but he is a reporter. I suggest we let each other know when we are contacted by him and what we say," said Clint.

They all shook their heads in agreement.

"Also, if you find any tracking devices, bugs, or taps, we let the others know. Use the email I will be setting up. Buy a separate phone to only use for communicating with each other. Here is my number," Michael said, giving them each a piece of paper. "I will go on and install my software on your work and personal computers. If you are being hacked, it will light up and tell you where it is coming from and ask you what you want to do. I will not invade your privacy, but those hack attempts will be relayed to me, and I will deal with them. So, just say "nothing" when asked what you want to do." They nodded.

"I will need you to add that software to the computer at this domain name. At this time, I will ask you not to ask about it or go there; just add your software," said Clint. Mary just looked at him, not knowing about the farm.

"All right, if that is what you want."

"In time, you will see why."

Michael left the grounds first, all of their paranoia at new heights.

"I take it there are things I don't know about," Mary said.

"Yes, a few, and the time is coming when you will need to make a choice about how deep you want to go."

"I've already made a choice. I am all in." Clint gave her a big long hug. They each went different directions. Clint decided to stay and visit the Cathedral.

12

Clint had two switches: on and off. He found it increasingly difficult to turn the switch off, even when he was with his children and grandchildren. His obsession with the positive and negative sides of gene-editing was taking over, and he had no idea how to stop it. There was no one to talk with about the dilemma other than those who were also caught in it. He would have to share the outlandish ideas he'd concocted to try to save the world from itself. They would lock him either in an asylum or deep underground dungeon never to be heard from again – alive or dead.

He'd considered just letting it all go. So, what if some country discovered a way to create a virus and hold the world hostage? He was in the last third of his life if genetics meant anything. Life had treated him well, and he was on the edge of producing something that would stun the world for the good. One last patent, one last farewell to humanity, and then just retire. The doctor laughed at himself. Sleep would disappear, anxiety would increase, and guilt would visit him daily in his waking and his sleeping.

Dreams had always been an important part of Clint's life. He kept a dream journal since his teens, and now there were over 30 journals filled with them. Believing they came from God or the unconscious made him pay attention. While he'd never admit to it, his dreams had led him to over half his patents. Genius often lay in the unconscious. Most people saw and acknowledged that reality in literature, art, and music. They balked when you talked about it in light of science. One of the many shortfalls of the scientific community.

Recently, the dreams had been one of two types. The first was of a more destructive nature. There were the overtly non-symbolic types where he dreamed of things overtaking the world and making humanity extinct. Life survived; humanity did not. Now and then, he was at the center of the dream, but more often than not, he was either helping or trying to stop the annihilation. There were dreams where he became the virus, not unlike Kafka's *Metamorphosis.*

Being a virus was a strange concept. You float around, doing nothing, having no purpose, and in essence, no life. Then all of a sudden, you come into contact with another entity, and you wake up. You attach yourself to that other "thing," and you come to life. Then your job is simply to multiply, and in so doing, you create a mammoth path of destruction – until you no longer exist. What an odd life. It was all happenstance. Viruses were not consciously seeking out hosts. Even in the dream, as a virus, he had no consciousness.

His most fearful dream was when General Fleming knocked on his door, had him arrested, and threw him into a dark dungeon, attaching things to his brain to pull out information. The General took the form of a virus, and when he'd acquired enough information, he completed his own metamorphosis and took over the world. He infected the world with himself. Clint always woke up at that moment. Inside, he knew how dangerous the General and his ideas were.

The meeting of him and Michael had come to him in a dream. Having never heard of his name or read any of his work, the dream revealed a paper to him. He read it in the dream. When he woke up, he ran to his computer, typed in the name of the article and the author, and there it was. The rest, as they say, is history.

One of the recurring dreams that scared him the most involved his children. They'd been attacked and threatened. They were pleading with him to do what their attackers were demanding. They had no idea what he was up to. They just kept begging him to stop. In one of the scenarios, communication simply stopped, and he couldn't reach them. He was at the lab, and the General was smiling. Save his family or save the world. The answer seemed so simple on paper.

For a long time, he'd chatted with a therapist about these dreams, but now they were too close to home, and his paranoia was rising.

Mary and Clint agreed there was nothing else for them at the conference, and they went home. They talked about next steps in their work mostly, and some about their spouses and kids.

Before they left New York, they stopped at the store Tom had mentioned. As expected, the owner was expecting them and took them to the back room. He gave them a duffle bag of things, explaining each one, but stating there were easy to follow instructions with each device. He also told them if they needed more information, to contact him.

13

A few months went by as the lab moved forward. There were several experiments going on, and they had to be careful they didn't mix them up. On any given day, six different projects needed something. Michael had been sending Clint some software that was proving quite helpful in speeding up the process of identifying segments of RNA and DNA. Clint thought to himself that Michael would have been a good epidemiologist, he picked things up quickly, and his mind was well outside the box.

One day, Clint went to Baltimore to talk with some colleagues about his work on malaria. He wanted to share some of his and Mary's findings. The process of bringing something to market was long and tedious. America was perhaps the slowest in the world. The regulations were almost unbearable and very expensive. During the Covid-19 pandemic, they got rid of many of the rules and ended up with many vaccines that cost billions of dollars that were essentially useless. They did find a couple that worked, but not above 74% efficacy. Clint was hoping that Michael's efforts would pay off on that front. His meeting with the people in Baltimore was a success. He'd come away with some new ideas to try and their commitment to being a testing site when he was ready. In the old days, the military used soldiers or prisoners to test new medications and vaccines. That was banished in the '70s. Now they had to find volunteers who knew what they were getting into. It slowed the process down, but human rights mattered to Clint.

Clint got in his car to drive home, and the phone rang.

"Dr. Williams?

"Yes, this is he, who is this?

"Bill, with UPS. We have a delivery from Incidyne and want to know where you want it?"

"Where are you?"

"South of Philadelphia."

"Do you have a phone I can call you back in ten minutes?"

"Yes." The driver gave him the number.

Clint found a phone booth and called back. "Hi, is this the driver, Bill?"

"Yes, sir."

"You have two large crates in your truck?"

"We have more than that, but two are yours."

"OK, take one to this address. I will meet you there and take you to the other address." He gave them the address of his farm.

"Yes sir, we should be there in an hour."

"I will meet you there."

Clint was lucky he was less than an hour away. He stepped on the gas and headed to the farm. On the way, he called his wife and said he'd be late. She was used to those calls. Luly didn't know about the farm. Clint didn't like keeping secrets, but it might be dangerous for her to know about this. He met the driver at the entrance to the driveway. It was a quarter-mile through the woods to his house and property. The driver pulled in front of the barn.

"If you could just put it around the side of the barn, that would be great."

"I'm not sure what is in this, but I know it is very expensive and fragile. Can I put it in the barn for you?"

"No, I have the equipment to do that, but I appreciate the offer."

The driver unloaded the machine.

"Now, where?" asked Bill.

"We need to go about two hours away, to Fort Detrick. You have three choices. I am happy with any of them. You can spend the night here, and we can get an early start, you can stop at a hotel somewhere between here and there, or we can just take it there now, and you are free to go."

Bill looked at his watch. It was 3:00. "If we go now, I can be home by 8. Let's do that."

"We can save some time if we take some back roads. They are paved."

"Not a problem with me. We will be in daylight, just keep me in your rearview mirror. If we get separated, I will just meet you at the gate to the base."

"Fair enough."

There was no easy or direct way from the farm to Fort Detrick. Clint found the drive relaxing and beautiful as it wove through farms, forests, villages, and the outskirts of cities. He found the names of the villages amusing. Dover, Camden, and Ruthsburg all had a British ring to them. There were Wyoming, Petersburg, Queenstown, and Stevensville. These were all before you went over the Chesapeake Bay Bridge. His farm was on the Delaware River that ended up going through Philadelphia. The Chesapeake Bay ended at the headwaters of the Susquehanna and the Elk Rivers. After skirting Annapolis, they would travel between Baltimore and Washington D.C. Over time, the two cities were becoming one. They went through Clarksville and got on 70, the major East-West freeway. If they stayed on 70, 2000 miles later Cove, UT. would appear. Two and a half hours later, they pulled into the Fort.

"Evening Sergeant, just making a delivery," said Clint to the man at the gate.

"Yes sir, have a good evening. How long will he be?"

"Twenty minutes."

"Thank you."

The gate opened, and they drove in. The base was fairly non-descript. It sat on the north edge of Frederick, near the Battelle National Biodefense Laboratory. Private and public industries worked hand in hand on many things. They parked in front of a two-story office building by the garage door. Clint went inside, and a minute later, the garage opened.

It was at moments like these that Clint was thankful for black budgets. When he'd taken over the lab, Level-4 was a total of 2,000 sq. ft. When dealing with decontamination, air filtration systems, and the testing equipment, there was only room for a few people at a time. Over the years, he kept increasing the size. It was now over 6,000 sq. ft. There was a space created just for the machine that was being delivered. He'd waited seven years for this day. The issue wasn't that it was backlogged; it didn't exist. When CRISPR came along, he'd worked with a company to build a machine that did the work of more than one hundred people in ten percent of the time. The fine-tuning of the mechanics and the software took two years. Now it was ready. The crate had been molded around the machine, so no parts moved. He knew there would need to be some adjustments, but he helped design and build the machine, so he was confident he and Mary could get it working correctly. Being from the government and offering them an extra million dollars to move to the front of the line and deliver it three months ahead of schedule was worth the price. Besides, it wasn't his money, at least not the one at the Fort.

The driver unloaded the machine and put it in a freight elevator at the back of the garage. Clint had thought decades ahead. He knew he'd need machines in his lab, and they'd have to be brought into the lab in pieces and disinfected. He had the elevator built with a variety of decontamination mechanisms. Once the new machine was in, Clint thanked the driver and sent him on his way with a hundred-dollar tip. He then sent Mary a text at 9:30 pm. "It arrived. See you tomorrow."

The pace of science, especially in the biological fields, was currently moving at a breakneck speed. Most scientists did all they could to live in the moment and not play catchup. Clint always lived five years out, at least. Part of his brain worked 25 years out. To him, that was not far enough. When he presented papers, which was not that often, he would leave the audience scratching their heads wondering what he was talking about. Many thought he was a quack until he produced a scientific miracle. Then they would all listen.

Earlier that evening, Mary arrived home. Her husband, James, and their young daughter, Samantha, greeted her with big hugs.

"Mommy," Samantha said, as she ran and jumped into her arms."

"How's my girl?"

She took her hand. "I have something to show you," said Samantha, giggling.

"I will be right back," Mary said to James as she walked down the hall.

Samantha was a budding artist. As they entered the kitchen, she ran to the table. The kitchen nook was surrounded by glass on three sides that looked onto their backyard. While they lived in a neighborhood, most lots were almost an acre. James loved to garden, and the yard was filled with flowers, butterflies, and hummingbirds. In the evenings, the fireflies came out.

"Look."

Mary was constantly amazed at Samantha's artwork. It seemed she could do almost anything. On the table that was covered with paintings and drawings, there was nowhere to put a cup of tea. She practiced her draftsmanship by cutting out pictures of magazines and copying them. On a scale of one to ten, she was already doing level seven work, and that was up from six two weeks ago. Then there the abstract pieces. Dazzling displays of color, geometry, and wisps of a child's mind.

"Wow, honey, these are amazing. You've been very busy the last couple of days."

"Yes, I have. Do you like them?"

"I love them."

"Which do you want?"

She was reaching for the abstract work when her eye caught another piece, mostly hidden under a drawing. She pulled it out. The drawing was much simpler like she would have drawn a year ago. She knew exactly what it was. She'd seen it in her dream. Mary sat down.

"I'd like this beautiful abstract piece, but can you tell me about this one?"

James was now seated with them.

"That was a dream I remembered. I wanted to draw it."

"What's it about?"

"Mommy, you should know what that is. It's the Capitol of America."

"What is on the Capitol building?"

"Some kind of giant bug is eating it."

"And who are those people near the building?"

"In my dream, they were running."

"From the bug, that would be a smart thing to do."

"No mommy, from the people chasing them, see?" she said, pointing to one group of people.

She looked closely and could see one small group of people running, being chased by another group that had guns. Mary and James owned no guns and tried to protect

Samantha from many of the frightening events occurring in the world. They felt seven was a little early to have to worry about that.

"Why are they being chased?"

"I don't know. They are good people. I think it's us."

"Was it a scary dream?" asked James.

Mary looked at him.

"Daddy, I already told you, it started scary, but then mommy showed up and killed the bug, and the people chasing us disappeared. I didn't like the bug or the people chasing us."

"Kind of like Scooby Do."

Samantha laughed. "Except this was real, not a cartoon."

"Do you have other dreams?"

"I have had this one a few times, and yes, I have a lot of dreams, mostly about playing or being a princess."

"Good."

James and Mary made a couple of drinks and went to sit on the deck.

"The garden is gorgeous."

"Thanks. I know we talk every night, but was it a good conference?"

"Yes, we learned some things I know would help with our work."

"What do you think about Samantha's dream?" asked James.

"I don't really know what to think. I rarely discuss work, especially not the scary parts."

"She does know you work with bugs that can hurt people, and that you are trying to stop them."

"I guess it makes sense she'd have that dream, but why being chased?"

"That part is a puzzle to me too. You do remember that she's had dreams in the past that came true to some degree. The healing of Tim, the death of aunt Sophie, and the bombing of the Lincoln Memorial, which fortunately failed."

"Those are hard to forget. Our daughter, the psychic."

"I don't want a bug eating the Capitol, and I don't need you saving the world. I don't want to brag, and I know it's God, but our daughter has some special gifts. We need to pay attention."

"Got it. I think we are safe." She gave him a kiss.

James was a botanist by training. He too had worked for the government until Samantha was born. He still did some consulting work, hoping to return to full-time work when Samantha was in high school. He primarily worked with seed banks.

The "Doomsday Vault," as it is aptly known, lies 800 miles south of the North Pole in Norway. It is buried into the earth and maintains a temperature of about -18° C. The idea behind the vault was to contain enough seeds to help humanity get back on its feet if there was massive destruction of the botanical infrastructure of the planet. Svalbard, it's actual name, can contain up to 2.3 billion seeds of 4.5 million crop varieties. New varieties were added each year. There was a worldwide committee that decided which of the hundreds of new varietals each year made it to the bank, and James was part of that group.

One of the concerns of botanists is genetic engineering. As new strains of plants are discovered that are drought and infestation resistant, fewer of the wild and natural varieties exist. Scientists know that most wild strains have existed for millions of years and slowly mutated to form new strains or hybrids. The same is true with animals. Wild salmon are slowly being replaced with farmed salmon. As they interbreed, the pure stock of wild salmon is gradually disappearing. Because of how easily and quickly pollen spreads, this same process is taking place on the plant level as well. Thus, the seed bank. The thought had not escaped either of their minds that this same strategy was at work in the world of viruses, although to date, it was all taking place in the natural world – other than perhaps in Samantha's dream. They'd not bothered to ask her where the bug came from.

Later, they put Samantha to bed and went out to watch the fireflies. The text from Clint arrived.

"What's that?" asked James, noticing her smile.

"Just Clint letting me know a machine we ordered just arrived."

"At night?"

"There are certain things Clint wants at any time. He's been waiting for this."

"Can you tell me what it is?"

"It's a machine that speeds up the process of gene-editing more than a hundred-fold in time, and the number of people it takes decreases dramatically."

"You aren't using this to create a giant bug, are you?"

"Of course not," Mary replied a bit nervously. That was not lost on James, but he let it pass.

"I wonder if the botanical world will get one?"

"They already exist at some level, and I have no doubt large botanical companies are using them to create new strains."

"Oh joy."

"I know you think we should slow the pace of progress down. I just don't see that happening."

"Think is the important word. I just wish we would spend some time "thinking" this through. The repercussions are potentially catastrophic for the human race."

"You forget to mention how some of the work will certainly save millions of lives," Mary said.

"I do tend to have my glass half empty while yours is half full."

"We need both," stated Mary.

They went to bed, cuddling, caressing the contours of each other's bodies, and making passionate love.

Mary looked him in the eye. "I will protect you from the bad bugs."

"What about from those people chasing us?" he said, smiling.

"Them too." They held each other as they dozed off. Neither of them would sleep well, their minds filled with different thoughts, most of which were less than pleasant.

14

Mary usually left the house around 7 am. They lived about thirty minutes from the base. She arrived and found Clint unpacking the crate.

The excitement on Clint's face was obvious.

"Looks like Christmas for you," said Mary.

"Better. I've waited years for this. We can finally get down to business."

The entire machine was about three feet tall by five feet long and four feet deep. It weighed about four hundred pounds. In the garage was a very small manual forklift. They moved it onto the elevator and stopped it at level 3. Here they would get off. On this floor, they would do their first change of clothes in a locker room whose air was separated from the rest of the lab. As they were changing, they'd pushed some buttons, and the first stages of the decontamination process had started for the machine. Millions, if not billions of microbes, were attached to them and the device. If those entered the air in Level 4, everything would be contaminated. Level 4 had the purest air on the planet.

The first step was to use UV rays. Everything in the air and on the outside of the machine was destroyed. The air was then heated to 200 degrees F. Most metals, plastics, and rubbers could handle that temperature. Clint had had the machine created with substances that could withstand 320 degrees. There were no completely closed areas of the machine. Clint was fastidious.

Mary and Clint went down a different elevator. They took off the scrubs they'd put on a floor above, took a shower, put on different scrubs, and then the specially designed suits that were completely self-contained outfits. In most labs, people wore very uncomfortable suits. They were cumbersome, and the gloves were clumsy. They couldn't eat, drink, or go to the bathroom. All this was to protect the wearer from the possibility of getting contaminated. Clint designed a version made of new materials the military had developed that fit more tightly to the body, had heating and cooling elements woven into the fabric, and had internal pockets for food and drink. They were able to pull an arm into the body from the sleeve, and the helmets were of such a design the hand could reach the mouth.

Knowing that after their work was done, they would need to shower and burn the suits, Clint devised a method of defecating in them, not unlike astronauts. They attached air hoses to the outfits that were connected to rails over their heads.

Airflow into a Level 4 facility was one-way. As you entered the lab, you almost felt pulled in as no air was allowed to escape other than through the air filtration system that lay under Level 4. The air was purified and circulated. Because the air they breathed was different than the air in the lab, there was no need to change the air in the lab. Level 4 labs were only built in areas with very little geologic movement. The new ones were built to withstand a 9.5 earthquake.

The final decontamination for the machine was to put it in the last airlock before the lab. A fine mist of highly toxic (to known bacteria and viruses) disinfectant was sprayed into the room. It was so dense the device was barely visible. They did this for thirty minutes. Finally, the air was sucked out, and pure warm air was put in to dry the machine.

Getting into this Level 4 lab took thirty-five minutes, getting out an hour.

An hour later, the machine was in its place and running.

"Let's do a test run," said Clint.

"Shouldn't we read the instruction manual?" asked Mary.

"Mary, I invented this machine, I know better than the people that built it what it does and how it does it."

She laughed. "Sorry, I forgot."

"Let's start by testing the malaria 24-b thru k protein strains. We will run thirty of each and then run the tests against the virus. Any that work with over 30% effectiveness, we will try on the mice."

The speaker in Clint's helmet talked.

"Clint, this is General Fleming. I am sending you a new recruit to train."

"Thank you for your generosity General, but we are just fine."

"I see you have a new machine. I'd like a report on that directly from you at the end of the day."

"I would be happy to do so. And again, we don't need anyone else down here."

"It's an order, not an option. You can either find room for her, or I will move Mary to a different unit, and you will work with Susan."

"Yes, sir."

"I appreciate your cooperation."

This now added another burden to Clint's load. He had to find things for the recruit to do that did not interfere with the work he and Mary were doing.

At one point, the General had told him they were wiring his helmet to broadcast everything that he was saying, for security reasons. Clint had threatened to resign. The General had backed off, but he was never sure he believed the General. While he was at the store in New York picking up the devices to find tracking and listening devices, he's spent some extra money and bought a very small one to use in the lab. Over the years, security had become a little lax with people like Clint. Everyone knew him; everyone trusted him. He put the sensor in his mouth during the shower and in his suit before going into the lab. He knew where the cameras were, so he turned his back, turned on the sensor, and raised it to his helmet.

A small beep hit his ear. The person at the store had told him the louder the beep, the closer you were. He started to walk around the lab, conscious he was probably being watched. He came to the place where the wiring for the computers and microphones was contained. The beep was loud.

He walked back to Mary and talked without talking. Mary banged his helmet, letting him know he couldn't hear. He wanted the General or anyone watching to think something was wrong with the microphones. This gave him an excuse to go back and look at the cabinet with the wiring. On top of everything else, Clint was an avid audio fanatic. He had one of the best sound systems on the planet at his home, which he'd installed. He looked at the wiring and moved things around. He flipped a switch that would only allow Mary to hear him. He would bring in a remote that he could let others hear if he wanted.

"The General has bugged the mics. He can hear everything we say. I took out the bug and changed the wiring. I will bring in a switch that will let us do that remotely. In the future, if either of us have something to say that is only meant for our ears, put an "X" on a piece of paper. I will flip the switch and let you know when we can talk without being heard. Oh, I almost forgot. The recruit must be a spy, so be careful what you say to her."

"Got it."

Clint went back and flipped the switch. "Can you hear me now?"

"Yes, all clear."

"Good." Clint was not worried about the General finding out they'd discovered his wiring. When Clint had confronted him and threatened to resign, he had the General sign a statement saying he would not "bug" the room. He'd broken the contract, and Clint was not averse to using it. The General would not admit he'd planted anything. Worst-case scenario, he'd plant another one.

The recruit entered twenty minutes later. Clint had thought the recruit might show up in a week or two. General Fleming had not said they were on their way.

"Good morning doctor, I am Captain Susan Fellows."

"Pleased to meet you Captain. I am Clint Williams, and this is Mary Soderstrum."

"Hello Mary."

"Hello Susan." Shaking hands in the suits was difficult. Usually, people just gave a slight bow.

"Susan, this is an awkward way of being introduced. I would like to spend a little time with you after work if that is possible, we will leave a bit earlier than normal," said Clint.

"I agree about the awkwardness, and that would be fine."

"Is this your first time in a Level 4?" asked Mary.

She laughed. "Ms. Soderstrum.."

"Mary."

"Mary. I am thirty-seven and have worked in Level 4 facilities for the past ten years. I have helped build and test two of them. I know this is probably the preeminent lab on the planet."

"We appreciate the compliment, and we certainly hope so," replied Clint.

"So, yes, I am very familiar with labs, the equipment in them, what they are used for, and the safety regulations."

"You are probably also used to having cameras on you. I am not sure you are used to having every word potentially listened to, not that you would ever say anything you didn't want others or the General to hear."

"I did not know that, and I appreciate the heads up."

"Please look around the familiarize yourself. There will probably be some things in here you have not seen or don't understand. If I have the time, I will explain it then, if not, then later when we meet outside."

"Thank you, doctor."

"Would you like to be called Captain?"

"I believe Susan will be fine."

"I am assuming you have Top Secret clearance."

"Yes, that has also been for the past ten years. The General gave me a briefing on some of the work you are doing."

For the next several hours, Mary attended to the new machine. They decided to call it Simon. In reality, its name was a Sylvester 28GD1000. Dr. Williams had designed what he wanted the software to do and had someone else do the programming. He wanted the computer to verbally tell them when each segment of the work was complete. When they were loading the software, it asked what they wanted to call it. After that was complete, it stated, "Simon says I am ready."

"How do I type in commands?" asked Mary.

"You shouldn't have to, just talk to it."

"Wow."

"What kind of a machine is that?" asked Susan.

"A very special one of a kind type for our work. Hopefully, you will get the chance to learn about it."

Clint showed her some work she could do under one of the four hoods they had where the bacteria and viruses are kept. She understood what he wanted and got to work.

The doctor left the lab early to meet with the General. He had to go through a decontamination process in the suit, put the suit, other than the helmet, in an incinerator, take a shower, go through another airlock, and finally up the elevator.

"Good afternoon General, how are you?"

"Sit."

"All business."

The General scoffed. "I want a report. You are not very communicative. Also, let's be clear. I am not happy with you doing things with or in the lab without my knowledge. I don't really care what you do or what you purchase. I just want to know. Like that machine you just put in there."

"That machine has sped up our work at least 40-fold.

"Impressive; where are the results?"

"I am hoping to have some significant results for malaria within the month and potentially for AIDS within the year."

"Wonderful news, now what about what I asked you to do?"

"General Fleming, no offense to your intelligence, but what you are asking, is for us to create a twenty-pound chicken without knowing how chickens are made."

"No offense taken, and no offense to you doctor, apparently other countries have been studying how chickens are made. We have intel showing that Russia, China, and North Korea are on the brink of creating a virus."

"I have my doubts. I don't doubt they want you to believe that."

"Don't underestimate the American intelligence community."

"Like Iraq's weapons of mass destruction?"

"I do not appreciate your insolence."

"And I don't appreciate your paranoia and mania about creating this new virus."

"Doctor, my concern is the defense of this country. Throughout history, when any nation acquires a new standard of weapon, they win. WWII and the atomic bomb is the most recent example. If you believe that powerful countries on the planet will hesitate to use a biological weapon, you are naïve."

"I understand that, I just don't want us to be that country."

"We are interested in defensive purposes only."

"Of course you are," Clint said, smirking.

"You have a job to do, do it. I need a time table."

"In this field, there are no real time tables. It's kind of like Edison with the light bulb. You keep trying things until you get it right."

"I want more time spent on the new virus than on fixing old problems."

"Like problems that could save over a half million lives a year? Like problems that could solve the disease that has cost more lives, conquered more empires, won more wars, and impacted more economies than any in the history of the world?"

"Yes, like those kinds of problems."

"I am doing what you ask, but if I can't do the work that is meaningful to me and the world, I will resign."

"Good doctor, perhaps I have not been clear. I don't like to make threats, but if I don't get what I want, your life will become problematic, not to mention the lives of those around you and those you care about."

"I like clarity General, good day."

He walked out of the room. Then he stopped the recording device and smiled. Two can play that game, he thought.

As he walked out of the building, two uniformed officers stopped him.

"Doctor Williams, please give us the recording device. You do know it is against the law to record conversations without permission," said one of the soldiers.

He handed it over to them. What choice did he have?

"The General is very concerned about you. We can see why." With that, they left.

Damn, he thought to himself. Not only did he have nothing on the General, they now had him breaking the law.

Clint was very good at keeping work and home separate. He and Luly had been married a long time and knew each other well. As hard as he might try, he kept little from her. Upon arriving home, he went out back to her studio, which sat by a pond. Hello love, how goes the work?"

"A good day. Some new ideas. And yours?"

"OK, thanks."

"You are lying, I can tell. What happened?"

"Words with the General, nothing significant."

"I get very frustrated with your work when I can't know what you are doing."

"I know, and I'm sorry. I'm not sure how much longer I can be there."

"Well, you don't have to leave," Luly stated.

"It's not that easy, Luly. Besides, what I have access to is the best there is."

"Is it worth it?" she asked, giving him a hug.

"I hope so."

She brought out a bottle of wine, and they sat drinking and chatting on a swing that swung over the water. The kids used to love flying of it when they went swimming.

"How is Tim?" he asked.

"I saw him today, took him out to lunch. Tim is Tim. He is so full of love. I think this would be a good weekend to have him stay over."

"That would be great. I will come home early Friday. I know how much he enjoys coming back here."

"He will never really understand why he can't be here all the time," said Luly.

"I know, but we both know it wouldn't work, and it's in his best interest. I ache for him."

"I know."

They held each other as the sun was setting, and the fireflies started to emerge. Clint saw them as ground-level stars flying in their own galaxies.

15

Michael had been busy himself. On top of overseeing the attacks on his machine and the aftermath, he was deluged with requests for the software, employment opportunities, letters and calls from rich people wanting to help him with a startup company, and requests for speaking engagements. Most of them, he ignored.

He was confident the network he'd set up with Clint and Mary was working well. Now he was sending them algorithms that helped their work, especially with the new machine. He'd found a way to get inside the software so he could just tell them what he'd done. Clint had the machine detached from the normal network and connected to Michaels. He'd told the General he was nervous about "outsiders" getting in. Simon, the computer, now had access to hundreds of millions of genomes, folded proteins within viruses, bacteria, and algae, and all the research being done. Michael had set his computer to find ways of making the software faster, broader, and deeper. Times had been cut in half already. His AI work was speeding up.

Michael was also spending time hacking into research labs in countries Clint was nervous about. While his software presented at the conference was defensive in nature, trying to protect computers from being hacked, he reversed the cycle and now was having the AI look for ways to break into machines. He used similar algorithms that used initial conditions, strange attractors, and isolating segments of programs that shifted. Upon discovering these, he'd put in his own piece of software that went undetected but allowed him easy access.

He used DARPA machines to tackle the AI work of self-organizing. Out of curiosity, he'd written a couple of programs that asked the computer to specifically let him know when pieces that related to the Butterfly Effect, Strange Attractors, and Fractals showed up. He was amazed at the frequency, especially around protein-folding. He believed biologic viruses were not alive, but they seemed to be somehow aware of how important initial conditions were. They had a type of gravitational energy that pulled RNA to a destructive tipping point, and he was finding deep patterns inherent within viral and bacterial DNA – better known as fractals.

Machine Learning could dissect an immense amount of data and find the needle in the haystack. If you knew what the needle looked like. Deep learning compared levels of information. Computers were asked to look for patterns on many different levels. These patterns were not unlike the Mandlebrot or other fractal sets. There was a geometry to programming. The programmer just had to change the lenses they wore.

The work was energizing to Michael. It was times like these he was happy not to be married. He was putting in eighteen-hour days and not thinking twice. He was believing more and more that he and the others could change the world. As he gazed at the monitor, showing him and the world know about people trying to hack in and the computer's response, he was amused.

Every time someone tried to hack in, the computer learned and adapted, not only defensively but offensively. It was like the computer was getting tired of the games, and the response became more and more brutal. Michael had to put in some boundaries. He wanted people to attempt to break the code, but he didn't want to completely destroy their computers.

There was a knock at his door. This puzzled him because there was a doorman that normally called to let him know who was there. In the door stood two people he assumed were some kind of federal agents. That's how they got through the door.

"Hello, gentlemen, to what do I owe this honor?"

"Are you Michael Longstreet?"

"I suppose I could say no, but then you know better."

"Being cooperative would be in your best interest."

"The firm, friendly type. Yes, I am Michael Longstreet," he said, changing his voice to be deeper.

"We need you to come with us, please."

I pulled out my phone and started to dial. One of them snatched it out of my hands.

"Hey, give that back, you can't do that. I'm in my own fucking house."

One of them flashed some kind of badge.

"Who are you with? I'd like to see that again. If you don't tell me who you are and why you are taking me, against my will, you will pay."

"I'm Mitch, this is Alex, we work for the National Security Agency. Mr. Spencer would like to speak with you."

"John Spencer? The head of the NSA?"

"That would be correct."

"Why?"

"That is above our pay grade. We were told to bring you in, willingly or unwillingly."

"Under what authority do you claim to have that right?"

He smiled. "National Security."

Sadly, the government could do pretty much whatever they wanted in the name of national security. In the last election, the incumbent president actually took boxes of paper ballots from polling places in states that were going to be close. He did it in the name of national security, claiming he had evidence of voter fraud. The Supreme Court was called into special session and mandated he either show proof or return them. They further stated that if there was any tampering with the boxes, he would be put in jail. The boxes were returned, he lost, handily, and left office.

"I was going to blame you, but I know you are pawns in their game. I am curious, do you have any boundaries? If they told you to kill me in the name of national security, would you do it?"

"Probably."

"In that case, I do blame you. Good luck having me on your bad side."

"We are not worried."

"You should be. Should I bring some things?"

" No, we were told you would be returned."

"How generous. No worries," Michael told the doorman. We got into the car and drove off. The CIA used to have an office in the World Trade Center. Since 9/11, they no longer broadcast where they were. The NSA didn't have a specific office in New York, so he was wondering where they were going. They ended up in the Lower West Side at the State Department Field Office.

They took Michael through the security check, up the elevator to the ninth floor, and into a small conference room.

"Can we get you something?"

"Martini? Shaken, not stirred."

"A sense of humor."

"I try. How about a diet soda?"

"I will see what I can do."

Michael was trying to figure out what this was all about. There was really no way they could know what he'd been doing, he'd covered his tracks well, and most of what he was doing was not connected to any network.

A few minutes later, John Spencer walked in and handed Michael a cola.

"Michael, I hope I can call you that. It's a pleasure to meet you."

"Wish I could say the same."

"I understand the circumstances were less than desirable."

"Abducting someone from their home is rarely desirable."

"My apologies. I really needed to see you, and I had hoped that the agents showing their identification and stating who it was that wanted to see you would do the trick. I guess not."

"They didn't show me their ID until I insisted, and they didn't mention you till well into the conversation."

"I apologize and will have them talked to. Sometimes they get a bit carried away. That was not the intention."

"Why am I here?"

"OK, right to the point. You have made quite a name for yourself in the world of AI, especially with that software you put on display at the conference a few months ago."

"I like to think I work hard and am good at what I do."

"Don't be humble; you might be the best in the world at what you do."

"I appreciate the compliment. What do you want?"

"I am sure you have been approached on all levels for your software. Obviously, the United States would like to work with you and reimburse you well for giving us the rights to that, exclusively."

"Yes, I have been approached. To be honest, I have not talked, texted, emailed, or otherwise communicated with another soul about it."

"I hope that is true."

"Look, Mr. Spencer. I was not born yesterday. I am certain you have hacked my phone and my email. You know what comes in and what goes out."

"Respectfully, Mr. Longstreet, and I do respect you, I have no doubt if you wanted to send a text or email without it being traced, you could."

"That is true. I will give you that."

"So, can we work with each other?"

"There are two issues that keep that from happening."

"I'm all ears. I'm sure we can work through them."

"I'm less than convinced. First, it's a work in progress. If you have read my resume, you know I don't release software until the bugs have been worked out."

"Have you found a bug in this one?"

"Not yet, but I'm giving it at least another month."

"We could help find them if they exist."

"Mr. Spencer, you know I monitor all activity in relation to that program. You also know I work for DARPA and do consulting with a few other government organizations. I bet you have a list, in that folder of yours, of what has happened to the machines you own that tried to hack me."

Spencer laughed and pulled out a piece of paper. "Yes, I have the list. Quite an accomplishment. We want it for our side."

"I don't plan on giving it to the enemy if that is what you are concerned about. You have seen my bank accounts, at least the ones I let you see. You know I don't need the money. I don't need or buy a lot of toys."

"Let me be frank. We will get the software."

"Or what? Throw me into a prison?"

"Nothing like that. You are aware that other companies and countries want your software as well."

"Yes, I'm aware of that."

"You have a family don't you? Parents, kids, grandkids."

"Look me in the eye Mr. Spencer. I do not take threats well. If I even think you are doing anything to or with anyone I know, you will wish you'd never talked to me."

"Is that a threat? I was merely speaking of what other countries might do. We would never do anything like that."

"John, I have seen files that exist that you will never lay eyes on. I know what our government is capable of. Put this on the record. Stay away or pay the price."

"Duly noted. Understand that I will do whatever it takes to protect this nation," said Director Spencer.

"From whom? Sometimes the enemy lives in the house."

"One more thing. What work are you doing with Dr. Williams?"

"Who?" Michael asked.

"Let's not play games."

"As you know, he is working on some gene-editing to try to figure out the cures to several diseases. We met at a conference to talk about how AI might help. I told him to let me know what he needed, and I would do what I can, which I will."

"So just working on things like malaria, for example."

"Yes, that is one of the diseases he is looking into."

"Have you been any help?"

"I have no idea. He does not understand computers, and I don't understand gene-editing. Our worlds are quite separate."

"He has not asked you to do other things?"

"Like what kind of things?"

"So, that means he has?"

"Stop trying to be clever, no he hasn't, but what kind of things?"

"It's a national security issue, but if he does, you will let us know, right?"

"Why wouldn't I? You can be trusted with all information, right?"

"Mr. Longstreet, I am the Director of the National Security Agency. I have the highest clearance in the land. I oversee sixteen agencies. Nothing happens without my knowledge."

"If you are truly that naïve, you are in the wrong job. First, your brain does not have the capacity to know everything. In the past year alone, there have been over 130 instances of spies, leaks, hacks, and abuses of power. I did my homework. If you knew everything, none of those should have happened. If you did know and they happened under your watch, you should be fired. I grow weary of the egos of the powerful. A little humility would suit you well. At least a reality check now and then. When all you do is listen to the "yes" people around you, bad things happen. Your job is to seek the truth, not your truth or that of the President, but the truth - it does exist."

"Are you done?"

"That felt good, and yes, I'm done."

"Then, we are done. I appreciate your coming in, even though you didn't have a choice. You will let us know if Dr. Williams asks you for anything outside the norm."

"Of course. Oh, one more thing. I am aware of your tapping my phones, hacking my computer, and having a tracker on my car. Unless you produce an order from a judge showing you have the right to do so and what evidence you have to get such a decree, things will happen. So, if you put a new bug on my car while I've been away, I suggest you have it removed."

"Another threat?"

"I told you I don't make threats." With that, Michael walked out of the office.

The conversation infuriated Michael. It was time to get serious.

Mr. Spencer had the tracking device removed.

16

Clint had been in the business for a long time. He knew the key players around the globe and enjoyed listening to them, reading their papers, and talking at conferences. He'd decided now was the time to move to the next step. He would be announcing his solution to the malaria problem within the month if everything stayed on track. The work that Michael was doing, along with Simon, had pushed his work forward by years. It also meant the work on the virus was progressing rapidly. Clint wanted to see what others were up to.

Clint flew to Germany for a conference. He had ensured that four of his colleagues from China and Russia were there. Their counterpart in North Korea could probably not be trusted. Most core technologies had come from the United States. We would get the ball rolling, and other countries would steal the technology and add to it. North Korea was a prime example. Their dictator had been focused on nuclear weapons, but the word on the street was he was now pouring money into genetic research. CRISPR technologies were available to anyone who had the understanding and funds to use it. The United States had put a ban on sending gene-editing technology to the countries I have listed. North Korea received most of their technology from China – for a price.

With companies formed the way they are now, it is difficult to find out who actually owns the information. Clint had no doubt North Korea owned tech companies that passed on information that was illegal to share. He would have to figure out another strategy for that nation.

He decided to meet with each person individually at the conference.

Chen was a scientist at the Wuhan lab that was the focus of the Covid-19 pandemic. She was in her early 70's, had worked at the lab for ten years, and was near the top of the list of people who understood CRISPR. She also let it be known that nations who attempted to create biological weapons were making a big mistake. Clint believed it was said sincerely. He wasn't sure what price she paid for saying it. Her English was better than his, having studied at Harvard for five years. At one of the sessions, he talked with her and handed her a note telling her where to meet him for lunch and to be careful not to be followed.

"Hello Chen, I'm glad we have a chance to talk."

"Your note about not being followed makes me a little nervous."

"I do not know how closely you are watched by your government, but mine has stepped up surveillance on me."

"Why?"

"Let's just say I have been asked to do something I do not want to do, and my guess is your government and scientists in other countries have been asked to do the same."

Chen looked away but did not answer.

"As I thought. I decided to take a risk in talking with you because I believe you are as against biological weapons as I am. The question is how to stop it without getting harmed or having others harmed."

She nodded.

"At some point, you will need to talk. Trust me, I am not wired. I'd be convicting myself if I was. I simply want to stop the madness."

"I agree."

"I am less than a year away."

"OMG, how? We are at least four, maybe six."

"I suspect the Russians and North Koreans are a bit behind you, but I could be wrong. I know people in Russia, but not in North Korea."

"I may know one. He is at the conference."

"Be very careful when you talk with him."

"He has already stated to me he is thinking of defecting."

"That's a good sign."

"So, what is the plan?" Chen asked.

Clint spelled out the plan to her in detail, with dates.

"Do you think you can do it and that it will work?"

"Everything would have to fall into place. Until the last moment, we can pull the plug, and no one would know."

"That is true. How do I contact you?"

"I know they are listening to my phone calls and watching my computer. Here is a number to call that they aren't tracing, but ensure you are on a line that is not attached to you. Here is an email you can use, and again, be careful where you send it from."

"Why do you think you are so close to creating this virus?"

"I have access to some new toys and to one of the geniuses on the planet in the area of AI. Between the two, we are progressing rapidly on several fronts."

"I assume you will not be sharing that information or technology."

"At this time, no, sorry. Within a month, things will start to happen. I will let you know, and you will see it on the news."

"I look forward to that. In the meantime, what do I do?"

"You need to start thinking about what we discussed and how you will pull that off."

"That will take some thought."

They both laughed, knowing for both of them it would be a struggle.

Clint met with the Russian scientist later with the same outcome. He received a consciously optimistic note from Chen the next day with regard to the North Korean. He liked it when things came together.

The doctor's paranoia was growing. He believed people knew he was at this conference, an international event. They probably knew he was meeting with Russian and Chinese scientists. In his mind, that brought on the attention of the CIA. He was correct. Since 9/11, the ODNI has helped pass information from one agency to another. Clint, Mary, and Michael's names kept popping up. They were on a lot of lists. And those lists had moles.

The next two months passed without any major event, other than Mary, and Clint found an answer to the malaria question that was 98% effective. With the help of AI, they created a benign virus that attaches to the liver. The folded protein in which it was wrapped activated the DNA when the malaria parasite moved to the liver, the starting point of its deadly lifecycle. Once it attacked the parasite, the inherent CRISPRS were tricked into folding the parasite's DNA in the same way. The parasite morphed into the attacker of the parasite.

They'd performed over a thousand trials on mice and pigs and had just completed tests on three hundred humans in Uganda. Using humans in other countries was far simpler than in America. Because the parasite was not the source of a potential pandemic, they could transfer the solution out of the lab with relative ease and complete secrecy.

Clint called Senator Brower and asked for a private session with the committee and the head of CDC.

"What is this about Dr. Williams?"

"I can't say on the phone, but the sooner the better, and I believe you will all be very pleased."

"When can you make it?"

"You name the time and place, and I will be there."

"Same room, tomorrow at 3 pm."

"I look forward to it."

"So do I, it better be worth it."

He'd not been this excited in years.

17

The following day he drove into D.C. with nine briefing books, each with four pages. One had two hundred pages. That was for the Director of the CDC who would want the evidence.

The meeting was called to order by the Senator. There were seven senators there. Two had said they did not have time for the doctor's nonsense.

"Madam Chair, I would like to just say that I hope this is worth our time and effort to come to this meeting. I have never been called on such short notice to a meeting where I do not know the purpose, other than in the case of a national emergency."

"Duly noted," Senator Brower said with some disgust. "Dr. Williams, proceed."

"Senators, if any of you would like to leave, please feel free. I come this time not because of a national emergency, but because of an international success. My team has developed the cure to malaria that will save over a half million lives a year. As you know, the mosquito that carries this parasite had decimated the planet since before humans appeared. Malaria has caused more deaths than any other disease or creature. Its end is now within reach." He pulled out a small vial and put it on the table. "Here it is. If some of you would like to leave, now would be a good time."

He stood and gave them each a briefing book, the thick one going to the Director of the CDC.

"Do not feel bad that your briefings are shorter than Dr. Cummins. He has all the data to show what I am saying is true."

"This is stunning and shocking news," said Senator Brower. "To the fine senator from Idaho, was it worth your coming?"

"I will admit that indeed it was, and I apologize for my outburst."

"We have not completed testing, but we should be able to do so within the month. Because this was developed in a government facility, the rights to the product are yours. Personally, when we divulge how it is produced, we hope it becomes public information. The government should not produce this, and we feel it unfair to give it to a company who can monopolize it and make billions at other's expense."

"We appreciate your concerns, but that is our decision."

"Senator, I am not stupid enough to bring the formula to this session. You could throw me into jail and send your experts into my lab, and they couldn't figure out how we have done what we have done. I see the Director looking at the briefing. I think he can tell you that there isn't enough information in that briefing to tell you how it's done, just enough to give him a sense of confidence we have indeed done it."

"Is that true Director Winthrop?" asked Senator Brower.

"Yes. What Dr. Williams has done is years ahead of what we are working on now. This is nothing short of miraculous."

"I will tell you that we are using the same science to discover ways of combating Ebola, AIDS, and the Coronaviruses that seem to continue to pop up."

"What do you want?"

"I do not need fortune. I have plenty. I do not need fame. I dislike what I already have. I will not share the technology we have developed because, in the wrong hands, things could go bad very fast. I want this vaccine produced in a minimum of factories and under my direct supervision, or at least under the supervision of the software I will put in the machines."

"And if we say no?"

"All evidence of this work will disappear. I will then go to a country that would like to do what I ask."

"Why so adamant, do you not trust the government you work for?"

"To put it mildly, no. I have seen no reason to trust the government on matters of urgency, and I believe the health of the world is an urgent matter. You could all keep this from humanity for years if not decades, bottled up in committee hearings, regulated out of existence, and holding many countries hostage. Not on my watch."

"Are you threatening us?"

"No Senator, those are promises. If you wish to challenge me, be my guest. I suggest you chat with the Director first and see what he thinks. He is a world authority on gene-editing. I see him smiling and nodding his head, he gets it."

"Dr. Williams, we deeply appreciate your bringing this to us. We will need some time to deliberate and come to a decision."

"I anticipated that statement, Senator. I do not trust, or like, General Fleming, whom I work for. Therefore, I will hand you my letter of resignation. You will see it is dated for three weeks from today. If I do not have the answer I want, that is what will happen. Just to let you know, if you attempt to come in and take over the lab and the work, I have a failsafe button that will destroy all of it. Do not test me. I know you are not often talked to in this way, but millions of lives are at stake, and I am not playing games."

"You are correct that I don't appreciate your tone or your words. I do understand. We will be in touch."

"One other word of caution. There are only twelve people that know about this other than one person in my lab. The people seated in this room. If word gets out, that means one of you committed treason. I will expect a full investigation and life in prison for that person. While that is taking place, I will release the formula to the world. You can put me in jail. It won't stop me."

"Dr. Williams, we all know how to keep a secret; we have top-secret clearance."

"Respectfully, Senators. The Hill is a sieve with holes so large a dog could walk through them. I ask you to trust me. We have done something no one else is close to doing. We have done it with some very advanced technology. All of your intelligence agencies may not find the leak, but I will. It will be one of you, and you will spend your life in prison. Again, not a threat, a promise. You know my past, and you know I am a man of my word."

"Yes, doctor, that we do. Your warnings are noted – but not on paper."

Everyone got up to leave.

"Stop, please," said the doctor.

"I am taking the briefing books with me. One of you accidentally leaves that on a desk or somewhere it might "inadvertently" be found, it's over. Please sit and read them, it won't take long. I will sit with the Director as long as she needs. I suspect she has already seen enough evidence to know what I am saying is true."

"I do not need to see anymore, but I do look forward to learning more about this discovery when you are ready."

"You will know almost everything there is to know."

Two Senators slammed the folders down and walked out. Six others took between fifteen and thirty minutes to read the four pages. Parts of it were well beyond their comprehension.   Eventually, they all handed him the briefing books and left.

"Clint, that was a bit rough, don't you think, especially for such good news?" Senator Brower asked.

"Senator, at times, I know that when people speak kindly, and with innuendo, the points are missed. I did not want anything missed today."

"You succeeded."

"I hope so."

"What does that mean?"

"To be honest, I don't believe they can keep a secret."

"With what is at stake, I hope so."

"Good day, Senator."

"Take care, Clint. We will be talking soon. I hope with good news."

He left. Driving home, he was elated. They were on the edge of monumental change that would, over time, save billions of lives. There was no amount of money that could buy that joy.

18

Clint took Luly out to dinner at the diner. There weren't any four-star restaurants in the area. Besides, they knew everyone there. They ordered ribs and corn on the cob. They always brought their own wine, admitting they were wine snobs. Seven dollar bottles of wine just didn't cut it for them. The owner had no problem with their bringing their own wine – they left big tips.

He shared with her what he had told the committee. She just stared in awe at him. "You've really done this? Of course you have, it's amazing." She leaned over and kissed him.

"You can't tell anyone, and I mean anyone, for two weeks," said Clint.

"I promise."

"Not even the kids."

"That will be hard, but OK," Luly said.

"I have my reasons."

"Of that, I have no doubt. I am so proud of you. Is this the biggest accomplishment of your life?"

"Other than the brilliance of marrying you and having kids, yes."

The next day he went to Fort Detrick and told the others about the briefing. The General was not pleased. Clint no longer cared.

Clint and Mary now included Susan in the malaria work but kept the other between themselves. The information they'd learned from the parasite and protein folding helped them find new directions in the other work. He was very excited and decided it was time to start working at the farm a few days a week. The malaria discovery gave him a good excuse to take a sabbatical or to quit.

He left the Fort and headed home. It was 4:00, quite early for him. He called Luly to say he was on his way. They owned land on the Monocacy River southeast of Frederick, about twenty minutes from the base. This was the site of the Battle of Monocacy in the summer of 1864 and was to be the most northern victory for the Confederate forces during the Civil War. Clint took the backroads. There were several ways to get home, and he got bored with the more direct route. He was going to buy Luly some flowers but remembered it was summer and their garden was full of them. He was heading south on Ballenger Road when a car pulled alongside him. Noticing the driver and passenger had masks on, he knew something was wrong.

Clint was not a spy, and doing daredevil driving was not his thing. He pulled over. There were two cars, a total of three people.

"Get out," said the masked man, who happened to have a gun.

"What do you want?"

"We want you, get out and get in the backseat of the first car. Give me your keys."

Clint handed over the keys, grabbed his satchel, and climbed in the backseat of the SUV.

"Who are you, and what do you want?"

"Shut up."

There were a lot of driveways and backroads off Ballenger. He watched them drive his car to one of them, pull off, and park it. They handed him back his keys. Then they returned to the other car. One of them got out to drive that one. At that point, they put a hood over him so he couldn't see where he was going.

Thankfully, the ride was less than an hour. He tried to pay attention to the turns and decided they'd made some turns to disorient him. He had no idea where he was. The vehicle turned onto a gravel road, drove what seemed like a quarter mile, and stopped.

"We are here."

Taking Clint out of the SUV, they took off the hood. Theirs were now off as well.

"You are Dr. Williams, correct?"

"Yes, but you could have asked that when you kidnapped me."

"We were confident we had the right person." He was a well-built man who looked to be ex-military, as did the other two. His voice sounded like that of a drill sergeant, but Clint could tell part of that was an act. Clint thought he was probably forty.

Clint looked around. They were in the middle of woods. There was probably half an acre of exposed land with a one-story log cabin sitting in the middle of it.

"This is your new home for a couple of days, Doctor. You are free to roam the house and yard. However, if you should choose to escape, you will find that most unpleasant." The second captor put two bracelets around his ankles. "These are similar to those you use with dogs. You go near the invisible fence, you get a shock. Our shock will make you unconscious. We suggest you don't try it." That man was in his late twenties. To Clint, this seemed like it might have been his first job. He was nervous.

The third person was in his fifties, very tall, bearded, and tried to avoid eye contact. He started roaming the property as soon as he was out of the car. They all carried weapons.

They'd been here before to set up, so they had nothing to take into the house.

"Make yourself at home, doctor. Your room is in the back on the left. The kitchen is full if you get thirsty or hungry."

"Who are you, and what do you want?"

"I think we will be getting a phone call shortly letting you know that."

Two hours later, the guard's cellphone rang. They'd taken Clint's phone and computer from him.

"This one is for you." They handed him the phone.

"Hello?"

"Dr. Williams, what a pleasure to speak with you."

"Who is this?"

"That does not matter at this juncture. What I want is you and the work you have done on the malaria vaccine."

Clint noticed an accent but couldn't place it.

"You can't have either of them."

"That is why you are being held captive. This can play out in one of several ways. The easy way is for you to just agree to do what we ask. The second is to do a bit of torture you won't enjoy. Then we move on to your lovely family. I know how much you care about them. Would you like the addresses of all of them? I don't like torturing little children, but trust me, I won't hold back."

"How dare you."

"Come, come, doctor, you are hardly in a position to chastise me."

"They will be looking for me."

"They will, and they won't find you. But they will find pieces of you in a couple of weeks, or pieces of your family. I'm going to give you twelve hours to think about it, and then the fun will begin. Think carefully."

The phone went dead.

"You are all animals. How can you do this?"

"Doctor, to be honest, I don't know who you are or what you have that someone wants. We just know that we get paid a great deal of money to do this."

"Normally, I would say that because you aren't wearing masks that I won't be leaving here alive. But if they want what I have, I have to." He thought for a moment. "My family, that's the guarantee."

"You are a fast learner. When that phone rings again in twelve hours, life will start to get painful. What is your life, and that of your families, worth to you?"

Clint walked outside.

Luly knew that Clint was punctual, if not early. She'd called and left messages as well as texted. No response. Then she called Mary to find out if he was still at work. Cell phones didn't work in the lab four stories underground. They had a landline that was patched through. No one answered. Then she called Mary.

"Hello?" answered Mary.

"Mary, this is Luly. Was Clint at the lab when you left?"

"No, he left well before me, probably around four."

"He hasn't come home, messaged, or called. I'm nervous."

"Let me make a couple of calls and see if I can find anything out," Mary said, trying to voice confidence and having none.

She hung up the phone and called Tom on the special cellphone.

"Hi, what's up?"

"I'm nervous. Clint left the lab three hours ago and has not arrived home or checked in. I think something has happened," said Mary.

"Hold on." I got my computer out and turned on his tracking device. I knew he had a tool that would discover trackers when they were on, so I purchased one I could turn on and off remotely. It wasn't cheap, but I had a feeling something like this could happen. I saw the blip on the screen, and it was nowhere near his home.

"Thanks for the call; I will take care of it from here. I will be sending you an address. If you do not hear from me by 10 am tomorrow, call the military police and tell them what has happened and give them this address. Do not, I repeat, do not call them until 10 am."

"Why not just call them and go in now?"

"My guess is that if he has been kidnapped, they don't care about his life versus theirs. You don't want negotiations, and you don't want a fight."

"What are you going to do?"

"On that, you will just have to trust me."

"In this case, that will be hard."

"I know, and you can call the military if you want, but you are risking Clint's life."

"All right, what about his wife?"

"I'd lie, but that might not work. We don't know for sure he's been kidnapped. I need to find out."

"OK."

19

They hung up. Mary called Luly back and convinced her that her husband was in part of the lab that didn't have a phone. He is very wrapped up with the malaria vaccine and maybe pulling an all-nighter. He'd done that many times, and while Luly was upset he hadn't called her, she breathed a sigh of relief.

I opened a locked box in my closet that was filled with phones. They were all coded, so I had to go into the password-protected folder and into the password-protected file to find the right one and the right number to call.

"Tom?"

"Yes, is this Ward?"

"Yes."

"I have a problem, and can use your help."

"When?"

"Three hours ago."

"What do you need?"

"There's been a kidnapping. I need him back, alive. I don't really care about the people who have him. I'd prefer them alive as well."

"Do you know where he is?"

"Yes."

"That helps. I won't ask how you know that. How many are there?"

"No idea, my guess is three to five. I mean, how many can it take to hold one person hostage?" I asked.

"How much money?"

"There hasn't been a ransom set."

Ward laughed. "No, Tom, for us?"

"Ahh, yes, mercenaries. This is a wealthy man, I can't tell you, but it will be more than worth your while. You just have to trust me."

"I do. Where and when?"

"Can you be at the airfield in Jersey in two hours?"

"I will call you if my team can't make it."

"I appreciate it."

"We appreciate you."

Ward had been a soldier who was falsely convicted of treason. I had been the one to prove his innocence and have the guilty parties, a colonel and a captain, convicted. I had the utmost appreciation for the military, but there were bad apples in every line of work. He was a very strong 8 with the Enneagram. He had a very strong will, a clear sense of vision and purpose, and his values were quite high and in place. His wing was a seven, which came in handy. He reminded me of the old TV show, MacGyver. He is very self-confident, decisive, and can be confrontational.

Two hours later, I was at the airfield. I'd put another emergency call into a helicopter pilot who agreed to fly us to West Virginia. We were standing by the chopper when a truck pulled up. Three people got out.

"Hello Tom," said Ward, giving me a hug.

"Good to see you."

"These are my friends, Alyssa and Sonya."

They were in camo and were each carrying a large duffel bag that looked heavy to me.

"What we are about to do never happened."

"Of course not," smiled Ward. The rest nodded in agreement.

"The captive is Dr. Clint Williams. He works at Fort Detrick and is about to announce a major discovery. I think someone has found that out and wants the discovery. I have no idea if it's a foreign government or a company. We do not want him or the discovery in anyone else's hands. He is a very wealthy man and will compensate you accordingly. If you are not successful, I will compensate you."

"Assuming you are alive," said Sonya.

"We don't go there," Ward said.

"Noted, and I have confidence this will be uneventful. They will be unprepared since it only happened five hours ago. We will be there in less than two hours. I have aerial pictures of where they are and where we will land," I said.

"You are positive he has been kidnapped. This is not a mistress in the woods?"

"He has no mistress, trust me. I can think of no other explanation, and it fits with what is happening."

"Is it more likely to be a foreign interest or industrial espionage?"

"That is hard to say. What he has is worth many billions of dollars and could sway the balance of power in the world, potentially."

"This guy is important," said Sonya.

"That is putting it mildly."

I knew they were all well-trained and fit soldiers, but they looked like most civilians. Each was attractive and did not have that stereotypical military look to them. I admitted to myself that I was surprised by women being part of the team, but I trusted Ward.

The copter took off and headed for West Virginia. West Virginia is an odd-shaped state. Where they were headed was a slice of the state wedged between Virginia and Maryland. The pilot had found a field about a mile away from where Clint was being held. They didn't want to get too close for fear of the captors hearing them. They were counting on the fact their guard would be low, and no one would be looking for them for at least twenty-four hours.

We flew in low and landed in a grass field. I told the pilot the plan. Taking our bearings, we headed to the house through the woods. There was little underbrush, so walking with flashlights was easy. Ward went in front, and I was in the middle.

About twenty minutes in, Ward stopped. He put on night vision gear. He could see the outside light of the house.

"There is one person on guard. I will go around the edge and try to take out the guard. Alyssa, set your scope up. If I fail to surprise him, take him out. I will wave my flashlight if I'm successful. You all come directly to the house. Make sure your silencer pistols are ready. Tom, you stay put. Here are binoculars. Keep watch on the front door. If I give you an SOS with the flashlight, you can come in. If you do not see it, call the pilot. Here is his number. Then call the FBI. It means we are in trouble."

"I'm not expecting that."

"Understood, I like to plan for every possible outcome."

"Got it. Good luck, and thanks."

"To be honest, it's good to be back at work. These are bad people, right?"

"Very bad."

"How are you going to explain this to the authorities?" asked Sonya.

"I'm working on that. Keep your masks on, and you won't have to worry about your part," I said.

"We have to be in and out before dawn. The second we are successful, I will flashlight you and then call the pilot, he will land in front of the house," Ward stated.

"What about the bad guys?" I asked.

"If they survive, that will be up to you."

20

Ward took off along the edge of the field with his night vision on. It was partly cloudy with a slight breeze. The rustling of the grass and the trees helped mask the slight sound of his movements. The guard stayed within a hundred feet of the house and had no night vision. They weren't expecting company. Walking at a slow pace, the guard, with an AK-47, circled the house. There was one light on in the living room. Their infrared sensors had spotted the others in the house, lying down. Those goggles were on Sonya. Once the outside person was taken care of and they were on the move, she'd be the point person for entry.

Deciding that being next to the house would be a better place to attack the guard from, Ward waited till the guard rounded a corner, then ran the sixty feet to one side of the house, the closest to the woods.

The goal was to not let the man yell or fire the weapon. Avoiding both was difficult. Timing was everything. This was Ward's lucky day. As the guard rounded the corner, he turned his back to Ward and lit a cigar. Ward smiled as he ran the twelve feet behind him and rapped his head with the butt of his gun. The man dropped. Ward put chloroform to his mouth to ensure he stayed out for at least a few minutes. He went to the front of the house and signaled the others. They started to move through the field.

Sonya stopped, pointed to Alyssa, who got on her knees and raised her rifle.

The man who was the leader came out the front door with his gun aimed at Ward.

"Drop the gun now."

Ward did as he was told.

"I'm glad I had to pee."

Just as he said that, a shot rang out, he dropped his gun and grabbed his leg. Ward picked up his gun and moved to the side. The front door flew open with the third man running out – not very well trained.

"Drop it there, or you will be dropped."

The man wisely did as he was told. Ward waved his hands and flashlight that all was clear. He waited for the others to bind the captors behind their backs and empty their weapons.

"Nice show on the leg."

"You said you wanted them alive, although I don't know why."

By the time I arrived, Clint was sitting on the porch.

"Tom? What are you doing here? How did you find me? Are you involved in this?"

"You have a good memory. No, Dr. Williams, we are your rescuers. Here is your phone. Call Luly and tell her you are fine and that you will explain when you get home. She thinks you did an overnighter at the lab."

He made the call, and while she was not pleased, she was happy he was OK. While he made that call, I called Mary and let her know everything was OK.

All three of the captors were now conscious.

"OK gentlemen, you have one minute to tell me who sent you, then we start to shoot and chop pieces off your body, probably like you were going to do with the doctor," I said.

"How did you know where to find us?" asked one of them.

"My secret. Now, answer my question," I stated with some authority.

"We don't have to say anything," said the boss.

"You are right."

I took one of their pistols and shot one of the others in the belly.

"My guess is it missed your kidney. You might bleed out. You might not. Probably depends on if we let someone know you are here."

The man screamed.

"I doubt that made it through the woods. I will grant that you picked a good spot," I said. "I ask again, and the first one who talks gets a pass on having another limb torn to shreds."

"We don't know. We just followed her instructions."

"A woman ordered this? I hope you rethink things like this in the future. Do you even know who this woman is?"

"No."

"Just following orders for a buck. Might I suggest you not take jobs like this in the future. Wait, you have no future. OK, Mr. Leader, any final words? By the way, coming onto the porch was not one of your brighter ideas. More like an amateur. Am I right?" I said, looking at Ward.

"Yes, we'd never do something that stupid."

"I know you won't believe me, but I don't know who hired us. It was a phone call. Money was put into a bank account. We were told where Dr. Williams would be at a certain time," said the leader.

"When were you told?"

"Yesterday."

"How did you find this place so fast?"

"Amazing what you can find with money. It's rented with cash for a month."

"What service were you each in?" asked Ward.

"Army," said the leader.

"Army," said the guard he'd taken out first.

"Marines," said the other.

"I respect that. I have no respect for you doing someone's dirty work where you have no idea who you are working for and what you are doing. I strongly suggest you rethink that while in prison," said Ward.

"Come on, man, we are just doing a job. I'm sure you do things like this."

"We did when we worked for the government, lots of black ops. No more. We only work for people we know and for causes we believe in."

"You are lucky."

"Yes, we are."

"So tell us, pro to pro, how did you find us so fast?" asked the leader.

"I will leave that up to you to figure out. What are your two top guesses off the top?"

"I'd say a microchip in Dr. Williams or a leak in the organization."

"Good guesses. Both wrong, but excellent guesses."

Sonya and Alyssa walked out of the house.

"We have everything.

"Take their guns."

They all moved out of ear range of the mercenaries.

"The doctor and I will take their SUV. Ward, you and the gals take the chopper. We will settle up tomorrow. Sonya. Alyssa, are you OK if Ward pays you off?" I asked.

"Yes."

"Doctor, without the work of these people, you or your family would be harmed or dead, of that I have no doubt. I told them they would be reimbursed. Am I right?" I asked.

"Of course, what do you want?" the doctor asked.

"I thought 40,000 apiece would be good," I stated. I knew their going rate.

"That seems cheap," said Clint.

"It is more than adequate for a night's work," said Ward. The women nodded.

"You risked your lives."

"I'm sure from your vantage point that was true. Because we found out so soon after you were taken, this was the easiest job we have ever done."

"Clint, we will have a long conversation on the way home. I would like to suggest that you use their services for protection for you and your family until all this is settled for at least the next two months. We will need to protect everyone involved and your family, 24/7. Can you do that, Ward?"

"Of course. Let me know how many we need to protect and where they are. I will send you a price."

"Perfect. It will need to start soon."

"Tomorrow soon enough?"

"Plan on watching ten people."

"What are you talking about? I don't have that many in my family," said Clint, surprised.

"I will explain in the car."

The helicopter landed far enough away. The captives couldn't read any of the writing. They loaded all the gear of the captors in their van.

"Thanks again. We will be talking. I just wish I knew who hired them," I commented.

"I do too, but I truly believe they don't know."

"I agree."

"I have an idea of how to find out," said Clint, smiling.

"How?" I inquired.

"In the car," said Clint, laughing as he walked to the SUV.

"He knows something we don't," said Ward.

"He knows a lot we don't," I replied.

"Are all of our tracks covered?" asked Ward.

"We think so," responded Alyssa.

"Do you need to lose the rifle?" I asked.

"Bullets are untraceable," stated Alyssa.

"That was a nice shot, by the way," said Clint.

"Thank you, I've been practicing," she said.

"Keep it up."

"I do."

21

Clint and I got in the SUV and started to drive off. The captives were securely tied; they wouldn't be going anywhere for at least six hours. They had given them each a sedative, which they gladly took rather than being shot again. The others headed for the chopper that took off just as dawn was breaking.

I stopped driving once we were out of sight of the house.

"Take out your phone. Call this number and report that you were kidnapped and escaped. It's the FBI. Tell them they are tied up and give the location. Here are the coordinates. Let them know the kidnappers are sedated, and two have been shot, not critically. They need to send a chopper to get them. If they doubt you, have them call General Fleming. Give them his direct number. I am sure he answers calls from you 24/7. Then call the General, tell him what I just told you to say and tell him you will see him in a few hours. Let him know that no one else knows what happened, and he expects the General to keep the secret. His wife believes he has been at the lab all night, and for now, he plans on keeping it that way. Got all that?"

"Yes." He made the calls. He told the FBI he would contact them tomorrow, and they could come to the base to interview him if they wanted. The General was a bit put off, but agreed.

As was expected, the General was called, and the chopper was sent.

"Now Clint, how do we find out who did this?"

"First, how did you find me, and who are you?"

"I will explain that I promise, but you first."

"The person who had me taken called me on their phone, this one," he said, picking up the captor's phone.

"He stated he would call back at 10 am, and if I didn't agree to what they wanted, they would torture me, then my family, and if necessary, start killing them. Isn't there a way we can trace that call? They will think I am still there."

"Yes, that can be done as long as they don't ask to speak with the leader."

I tried to think three steps ahead. If they came after Clint, it was only a matter of time before they connected the dots and tried to do the same with Michael and Mary.

I pulled out another phone, Michaels. No one answered, so I left a message, frustrated. Wasn't everyone supposed to have that phone on and ready at all times? He laughed at himself.

"OK, now I will tell you what you wanted to know. I have been tracking you off and on since you went to Boston months ago. As you probably know, I write about bad people doing bad things. That all takes time and research. I believed and still do, by the way, that you want to do something bad. My problem is that you are a good person. With your plan, you needed help on several levels. One was human. You found Michael to help with the AI that sped up your work. Then you needed hardware because humans are way too slow. You designed and bought two machines. One is at Fort Detrick, and the other is at your farm."

Clint's eyebrows raised.

"Yes, I know about the farm. I am good at what I do. After I told you and the others where to buy equipment that tracked you, I had to change devices. That is how I found you, the device in your satchel. I had to put it in the lining, which by the way, isn't easy since that rarely leaves your side."

"It's hard to be angry with you about tracking me when you saved my life. What do you think I am doing that is so bad?"

"Creating a virus or bug that can destroy the world."

There was a long silence.

"You are both right and wrong. I do not know if I am a good or a bad person. I hope the prior. Yes, we are creating a new virus. I have no interest in destroying the world, but I am concerned that many nations and people do, including our own. General Fleming, who you know about, has ordered us to create a virus that can be weaponized. We created a solution to the malaria issue that has plagued humanity since our origins. He has pulled the plug on all research until we develop the virus. Michael's help has sped up the process of both up by years, if not decades."

"So, what do you do when you have this bug?"

"I'm not sure. I think I will hold the world hostage."

"People must know you'd never let it loose."

"First, we will have a vaccine before the virus. After that, I haven't thought it through to conclusion."

"You might want to do that."

"Are you going to turn me in?"

"I haven't thought it through. So far, you have done nothing illegal or wrong. I know that Mary and Michael are part of this, who else?"

"The General knows about the virus. Other than that, no one. I have two people, one in Russian and one in China, maybe one in North Korea, that are willing to help. They are waiting to see what I come up with."

"What is the end game? What do you want?"

"I want a treaty to ban the creation of new viruses."

"Is that even possible?"

"I'm not sure, but I have to try, at least to slow it down."

"A worthy cause."

"I like to think so. Perhaps it is folly. Will you stop me?"

"I will have to give that some serious thought. At this time, how close are other countries?"

"Probably five years or less. You need to understand something. We all know the world has enough nuclear weapons to kill most humans. But we also know when a weapon is fired, we have time to respond defensively or offensively. Each weapon is contained in a specific area. You can drop a nuclear weapon on Manhattan and annihilate it. That bomb does not physically or biologically affect Pennsylvania or Los Angeles. On the other hand, take a virus. We have seen what happens with the pandemic of 1918 and of 2020. Viruses spread like wildfire. What if you create a virus that spreads faster, and instead of a mortality rate of 1-5% like Coivd-19, it is 90%. Instead of an infection rate of 1-3%, it is 85% or higher."

"That's not possible."

"You don't know the pace of technology. It is possible. In fact, there is about a 0.1% difference between your genes and anyone else. We will get to a point where we can create medications and viruses that are specific to a human being."

"That is frightening."

"Indeed. I want to stop the train."

"And what is the cost?"

"Financially, not much. If everything goes well, no human life. That depends on the decision-makers."

"What if your bug or someone else's gets loose?"

"As I said, the virus I will create will have a vaccine before it's made. If it were to be exposed to the public, we simply vaccinate everyone."

"I shouldn't, but I trust your judgment on that one."

"I appreciate the vote of confidence."

"I don't have enough money to protect you and your family, let alone the others involved."

Clint laughed. "Trust me Tom. Money is not an issue. Do whatever it takes. I will set up an account with several million dollars in it that you will have access to. Does that work?"

"As long as Michael helps with that process."

"He is the one I am going to ask."

I took a minute to let this all sink in for both of us. This was, after all, our first real conversation. I was reading Clint as a 5/4 with the Enneagram. Highly intelligent, a research addict, logical, with an emotional and creative side to him. My research had shown he had a special needs son, and there were times that led to bouts of depression for him. He loved his son, saw him often, and when he spoke of him, it was always in glowing terms. While he seemed a bit anxious at the moment, I wasn't holding that against him. He'd just been kidnapped, was being driven by someone he didn't know, and was talking openly and calmly. I'd say he was pretty healthy.

"I've been doing a lot of reading about genetic engineering. I want to make sure I get things straight," I said.

"All right."

"I know it all starts with DNA."

"And RNA, as Mary would insist - correctly."

"Right. DNA is made of base pairs of GCA and T. I know when DNA splits, it forms a new strand with the right base pairs. But how are those things, I think they are called nucleotides, made?"

"You are getting into waters that are far more complicated than you need to understand. I say that respectfully, but people spend years on these things that we still barely understand. The human body is a miracle. DNA is a miracle within a miracle. We take it for granted, but the number of things that have to go right is astronomical. In fact, mutations are constantly occurring;

they just don't all last. We don't understand why some do and some don't."

"OK, so let me know the important parts."

"The nucleotides in DNA are made up of three parts; a nitrogen base, five sugars, and a phosphate group. The letters you mentioned are the four nitrogenous bases. There is a fifth that is found in RNA, U, instead of the T in DNA."

"Sounds like plant fertilizer."

"You catch on quickly. DNA has been around since the beginning. Simple plants have much simpler DNA. DNA is DNA, a plants is no different than a humans. The same ingredients are in all of them. The difference is in the ordering of the nucleotides and how they are expressed. The groups of nucleotides form genes or segments of DNA. These are chromosomes. Humans have 46 chromosomes, and plants have 24. Proteins, called histones, link with the DNA forming chromatin, which then makes a chromosome. One DNA cell is simply a long strand of nucleotides – up to six feet long if stretched out. There are about three billion nucleotides in that strand. It is the proteins that crunch them into a size that fits inside a cell. Genes carry the instructions for how to make the proteins that do the work of making us walk, talk, think, and tick."

"The DNA sends the appropriate strand to the RNA, which sets up the recipe on how to construct the protein. The mRNA goes into the cell and finds a ribosome, which is the protein factory. We both consume and create proteins that, in turn, break down and provide the chemicals to make more. Everything in you is made of proteins, well, mostly. We have come to understand that the folding of proteins at every level impacts the outcomes. Right now, our work is focused on that. Because most algae are harmless and certain viruses attach easily to certain forms of algae, that is our work for future things. We are focusing on Phycodnaviridae. It has a large two-stranded DNA. In 2014 scientists believed certain strains of the virus could infect humans. Those are the ones we are working with. Also, algae are easy and fast to grow, unlike other types of viral substrates. All make sense?"

"I just received a semester-long course in genetics in ten minutes. Some things clicked, but it will take time."

"Don't get stuck in the details. I know people that read books and believe they understand it all. Trust me, they don't, and most would fail even a simple test."

"I agree."

"Are you going to write a book about all this? Will I be quoted? I know you like sending people to jail. Is that my fate?

"I can't speak too much of that at this time. I won't lie to you. I suspect there will be a book. My guess is that people will be going to jail, I hope not you. Yes, I will quote you, but I will give you a chance to challenge any part of the book, or I will at least acknowledge the discrepancies between your version and mine. Is that fair?"

"I will leave this little chart with you." The doctor reached in his satchel and pulled out a piece of paper. "This shows you the twenty core amino acids and how they are made. You will see that each has three base pairs. We call that a codon. Genetic engineering is broken down into codons. How and where these are lined up in the DNA strand makes all the difference. Somehow, we haven't quite figured it out yet. The body, actually the cell, knows which segments to choose to make and fold a protein. While all the DNA in your body is essentially the same, somehow cells know how to make blood, bones, muscle, organs, neurons, and the other 50 trillion cells in your body. We do not understand that mechanism. Remember, you started as one cell with DNA. Somehow as they split into what becomes you, bodies and plants know how to make different proteins for the different functions. Red blood cells have no DNA. Sperm and eggs have half the DNA, waiting for the other half from the mate."

"OK, my brain is on overload. Usually, I can take notes when I am listening or reading."

Clint laughed. "As you can probably tell, I love this stuff. Here's the chart for later. "

"I'd noticed. Do you always carry diagrams around with you?"

"They help my memory and meet with people all the time who need diagrams to help simplify issues."

Amino-acid biochemical properties	Nonpolar	Polar	Basic	Acidic		Termination: stop codon		

Standard genetic code

1st base	2nd base							3rd base
	T		C		A		G	
T	TTT	(Phe/F) Phenylalanine	TCT	(Ser/S) Serine	TAT	(Tyr/Y) Tyrosine	TGT	(Cys/C) Cysteine	T
	TTC		TCC		TAC		TGC		C
	TTA		TCA		TAA	Stop (Ochre)[B]	TGA	Stop (Opal)[B]	A
	TTG[A]		TCG		TAG	Stop (Amber)[B]	TGG	(Trp/W) Tryptophan	G
C	CTT	(Leu/L) Leucine	CCT	(Pro/P) Proline	CAT	(His/H) Histidine	CGT	(Arg/R) Arginine	T
	CTC		CCC		CAC		CGC		C
	CTA		CCA		CAA	(Gln/Q) Glutamine	CGA		A
	CTG[A]		CCG		CAG		CGG		G
A	ATT	(Ile/I) Isoleucine	ACT	(Thr/T) Threonine	AAT	(Asn/N) Asparagine	AGT	(Ser/S) Serine	T
	ATC		ACC		AAC		AGC		C
	ATA		ACA		AAA	(Lys/K) Lysine	AGA	(Arg/R) Arginine	A
	ATG[A]	(Met/M) Methionine	ACG		AAG		AGG		G
G	GTT	(Val/V) Valine	GCT	(Ala/A) Alanine	GAT	(Asp/D) Aspartic acid	GGT	(Gly/G) Glycine	T
	GTC		GCC		GAC		GGC		C
	GTA		GCA		GAA	(Glu/E) Glutamic acid	GGA		A
	GTG		GCG		GAG		GGG		G

Wikipedia: DNA Codon Table

The phone rang.
"Hello?" I answered.
"Michael. "

"Did you get my message?"

"Yes, and I can do it, but do you have another phone you can give to the doctor he can throw away?"

"Of course I do, you know me better than that."

"Have him call 129-399-5555," said Michael.

"There are no area codes that begin with 1 or 0."

"Trust me."

I looked in my pack and handed Clint a new phone and the number. He looked puzzled but dialed the number. Michael picked up and gave him instructions. Clint hung up.

"Wow, that guy is smart."

"Tell me about it. What did he tell you to do?"

"He said this number is untraceable because it doesn't exist. He will route the old number, the captor's number through this one, and through the cell tower near where we were. That way, the person on the other end will think we are at that farm. Michael said he doesn't think she will call from a number that is traceable to a person. It will either be a throwaway phone like these or a more public phone."

"Makes sense, in an odd kind of way," I said.

"I guess."

"So, you know what to do?"

"Yes. I need to keep her on the line for at least a minute."

"It used to be three."

"We are talking about Michael here."

"True."

"He said he'd call you as soon as he knows something."

I pulled into the road where the doctor's car was. We found it a few hundred yards in.

"How did you know where to find this? Is it bugged too?" Clint asked.

"Not currently, at least that I know of, but we will check. I pointed a gun at the undamaged person's heart and told him I wanted to know the location of the car. It didn't take long. He knew I'd pull the trigger," I said.

"Would you have?"

"Probably not, but if I wimped out, the people I hired, would. That is why they make the big bucks."

"Speaking of which, where did you learn to shoot? I mean, that shot to the stomach was risky.

"Let's just say the military trained me well."

"What do I owe you? You saved my life and perhaps that of my family."

"There is no fee. If I write a book, I will get the proceeds – all of them."

"I don't need them or want them, but do I get to preview it to correct mistakes — especially about gene-editing?"

"Of course. I will ask that you grant me some latitude on the storyline. I may make this encounter sound a bit more dramatic than it was."

The doctor laughed. "Of course, although to me, it was more than dramatic enough."

"Readers like more and more gore. I might have six people holding you hostage, and three die in a gruesome way."

"And if people ask, I will simply say what politicians say, 'I will neither confirm nor deny that.'"

"Agreed. You do know that I don't need your permission to write this book."

"Yes, I understand, and I also understand Michael could probably make it disappear." They both laughed.

"That thought had crossed my mind. OK, here we are. Are you sure you don't want me to follow you home and be with you on the call?"

"Yes, but thank you for all you have done."

"You have the phone I gave you for the call. When done, smash it and throw it away. Here is another one. Only use it to call me when you need to. For the next several months, I encourage you to keep it close."

"I will certainly do that."

"If you press 1, it calls me. If you press 5, it calls me and tells me you are in trouble."

"Thanks."

Clint got out, they gave each other a hug, and Clint drove off.

22

At 10 am, the call came.

"Hello, who is this?" asked Clint. He was sitting in his living room. He'd spent a couple of hours with Luly, calming her down after she got mad at him for being kidnapped. She laughed at herself for saying such a thing. He didn't want her in the room because he feared she might grab the phone, reach through it, and throttle the person on the other end.

"At this moment, that is of no concern. Have you reached a decision?"

"That depends on the answer to a few questions."

"I will make no guarantee I will answer them, but you also know I hold all the cards right now."

"No, you hold my life and that of my family. I hold the key to potentially billions of lives and dollars. I want some assurances if I agree that the vaccine will get to those in need, no matter how poor."

"On that, we agree, doctor. I want to save people's lives."

"And make money."

"It is not all about the money, but yes, we will make some money."

"I also need to know my family will be protected."

"If you agree, no harm will come to anyone. We have built in some assurances if you happen to tell."

"In other words, my family and I are permanently expendable."

"Well put. On the other hand, you will be richly rewarded for your contribution."

"You are so generous."

"Do I detect some sarcasm?" said the voice.

"Just a bit."

"I need an answer."

"Since I know you have assurances, I will go the next step. What do I have to do?"

"My associate will bring you a computer. You send the formula to an email I will give you."

"Wow, you need to talk to some scientists. You clearly aren't one. It doesn't work that way. First, it's in a computer at Fort Detrick. Those computers are encrypted and have the highest level of security. You can't send anything in or out. That's how we keep people from stealing things. I will need to get to the base and find a way. There would be at least two pages of instruction," said Clint.

"Can you send me enough to let my scientists know you have the answer?"

"Yes, I think I can convince them of that."

"Then let me talk to my associate. He will give you a computer, and you can send me what you can. Then I will be in touch. I will have him stay with you until you get me what I want."

"All right, hold on."

Clint went out on the patio, prayed that Michael had what he needed to trace the call, put the phone down, and dropped a large rock on it. The phone went dead on the other end. That person thought perhaps the cell coverage was bad.

23

Forty minutes later, Michael called me.

"I have two pieces of information. You must guarantee no one will find out how you got this."

"I promise unless I have your permission," I said.

"Fair enough. First, the mole is Senator Nelson from Tennessee. Second, the kidnapper is Emily Wallenstein, the CEO of Biosolutions, a startup vaccine company in Tennessee. She has had two others that were flops. The company is leveraged to the max and will probably be going under," said Michael.

"Do I want to know how you know this?"

"Briefly. I recorded the call. I traced the phone to Biosolutions headquarters. I figured it was someone high up. Her arrogance is misplaced. She didn't use a voice disguiser. I recorded the call, analyzed it, and compared it to some speeches she'd made. A perfect match. Then I checked the phone records of everyone on the Senate Committee. I started with Nelson since he's from Tennessee. Wallenstein has made significant contributions to his campaigns. There was also a phone call from his cellphone to hers an hour after the hearing. Finally, he opened a bank account under another name but again, stupidly, had the email address of that account linked to his personal Gmail account. Connect the dots and follow the money. He's the guy," Michael said.

"I know not to mess with you. I appreciate this."

"Thanks."

"They are moving to phase two, whatever that is. Perhaps you know. I am nervous about everyone's security. Clint is paying to have people watch all of you and your family. We will let you know if we get concerned, and you let us know if you get concerned."

"I don't like this, and by the way, take me off Clint's payroll. I will take the security, but I can afford to pay for it myself."

"All right. You do know you can pull out of whatever it is at any time."

"I know, but what might happen if I do pull out is something I fear more."

"Agreed. Take care."

"You too."

I called Senator Brower and left a message on her phone. "This is Tom Armstrong. I don't care where you are or what you are doing, you have one hour to call me, or I go to the press with something explosive."

Then I called her office in D.C.

"Hello, this is Senator Brower's office."

"Hello, my name is Tom Armstrong. Do you know who I am?"

"Yes, sir, of course."

"I left a message on Senator Brower's cellphone. She isn't taking calls."

"She is in a hearing Mr. Armstrong."

"Listen carefully. She has one hour to call me. If she doesn't, something explosive will be released to the press with full knowledge she had the time to respond. I suggest you get that message to her now. If you don't, kiss your job goodbye, and probably her career. Do I need to repeat any of that?" I said.

"No, sir."

"Good, the clock is ticking."

Thirty-five minutes later, the phone rang.

"Hello, Senator," I said.

"This had better be good, Tom. I am not amused."

"Are you amused by the fact that a member of your committee has committed treason and is part of a kidnapping plot?"

"What are you talking about?" asked Senator Brower.

"I have proof that one hour after your committee meeting, Senator Nelson called the CEO of a pharmaceutical company in Tennessee and told her the news. The next day, that CEO had Dr. Williams kidnapped. The FBI has the kidnappers, but not the CEO yet. They will, that is my next call."

"I need to check this out. How are you involved?"

"That is my business. I am sending you an encrypted email to your personal address that you gave me. The information you need is there. I will give you three hours to have him arrested for treason. If I do not see on national television in three hours that he has been arrested, I go to the press with everything. I think you know your career would be over. A Senator knowing about a kidnapping and doing nothing."

"Are you threatening me?"

"I don't make threats. I do not take kindly to someone who the American people are supposed to trust, betraying that trust or having someone kidnapped who is about to save the lives of billions of people. You can take all the credit if you want and win another term, or you can tend your garden. There are no other options."

"I can't just arrest someone on what you say."

"I suggest you call the FBI, or whoever you want, and check my facts. They have the resources to do it."

"How did you get all this information, and what about Dr. Williams?"

"He is fine and at home. He escaped," I said.

"How?

"You can ask him?"

"He got away without your help?"

"Ask him. I suspect the FBI already is."

"What about the captors?" she asked.

"As I understand it, he subdued them."

"Tom, come now. I am not that stupid," said Senator Brower.

"You have what I will tell you for now. My sources are confidential, and if you choose to push me on them, that would not be a wise decision," I replied.

"Another threat.

"Again, Senator, I don't make threats. The clock is ticking. Oh, I almost forgot. As promised, the committee broke theirs, so Dr. Williams will be handing over the formula for the malaria vaccine to the CDC. If they do not make it public within two weeks, he will." I hung up.

As far as politicians go, Brower was as good as they got, and that wasn't saying much. They all lied, and they were all self-centered and puppets. I had to admit to myself that I enjoyed taking them down. I also liked giving the good one's information that made them look good.

Next, I called the FBI in Baltimore and was connected to the agent overseeing the Williams investigation.

"Agent Sayer here, to whom am I speaking, please?"

"Hello Agent Sayer, my name is Tom Armstrong, do not write that down. I am more aware than you of what is going on with Mr. William's kidnapping. I know you have three people you are questioning. They know nothing about who hired them. I do."

"And how would you know that Mr. Armstrong, other than being good at your job?"

"We will leave it at that for the moment. Please give me an email address where I can send some encrypted information." She gave it to him. "What is your first name Agent Sayer?"

"Glenda."

"OK, Glenda. In less than three hours, Senator Nelson will be arrested for treason – sharing Top Secret information with the CEO of a biopharmaceutical company in Tennessee, Biosolutions. This CEO, Emily Wallenstein, paid off the senator and had Williams kidnapped. I am sending you all the proof you need. Like the person I gave a three-hour deadline to arrest Nelson, you have three hours to have Ms. Wallenstein arrested. That is more than enough time to check my facts."

"And if we take more time?"

"I will go public, you will lose your job when it is known you knew and did nothing, and Ms. Wallenstein may have fled the country. It's your choice."

"The FBI does not take kindly to threats."

"I don't really care, Agent Sayer. Go ahead, take your time. Let's see how that goes. The clock is ticking. By the way, if I find out you have told anyone - I mean anyone - how you got this information, my investigations into you and the FBI will be long and deep. Don't test me." I hung up.

I sat down with a beer and turned on the television. Everything was normal for two and a half hours. Politics as usual. Even Senator Nelson was upbeat, bragging about how he'd squashed a bill that would have regulated the cleanliness of the water in his state, reeling from a recent finding of lead.

Ten minutes later, the cameras were in front of his office, watching as he was taken away in handcuffs.

"I'm innocent," he screamed.

Senator Brower greeted the reporters.

"I will make a short statement. Senator Nelson has been arrested on charges of treason by sharing Top Secret information and for being an accessory to a kidnapping, which you will hear more about briefly, I'm sure. That is all for the moment." She looked at the camera. "Good enough?" I knew that was for me.

Yes, it is, I said to myself. Twenty minutes later, a reporter broke into the news.

"This is James Broderick reporting from Chattanooga, Tennessee. The FBI has just reported they have arrested the CEO of Biosolutions, a pharmaceutical company headquartered here. Emily Wallenstein was arrested on kidnapping charges. We do not know the details other than she organized the kidnapping of Dr. Clint Williams, a world-renowned geneticist. Three people are also being held in connection with the kidnapping. Dr. Williams is apparently fine. We do not know the details. Back to you."

The anchor guessed the two stories were connected. Quite a twenty-four hours.

The next two weeks were a whirlwind. Clint, Michael, and Mary had to tell their families what was going on. Some of those sessions were tense. They were all a bit fearful and relieved when the security started to arrive. Ward and his troops were in place within two days of the kidnapping. Devices had been placed in and outside of the houses to give warning. Each person was given a panic button if something was happening, wherever they were. Clint had an interesting interview with the FBI in Baltimore. They'd wanted to come to him, but he insisted on going to Baltimore. It gave him a good excuse to get to the farm and check on things.

24

"Welcome, Dr. Williams, come in please. I am Agent Sayer."

Clint had decided to let them keep the formal name. They took him to a conference room. There was one other agent in the room at the time.

"We simply want to debrief the recent events of your life."

"You mean the kidnapping."

"Yes." Agent Sayer preferred using calmer language with victims and harsh language with the perpetrators. Clint did not have time for games.

"What would you like to know?"

"Just take us through what happened."

"As I have said before. I was driving home when two cars pulled me over. At gunpoint, they put me in their car and drove off. My car was left by the side of the road. They put a mask over my head, so I could not see where we were going. I tried to remember distances and directions but gave that up quickly. There were too many turns. They wore masks. When we arrived at the site, probably an hour or more from where I was taken, they took off my mask and put the ankle bracelets on me to keep me contained. There were three of them. They received a call, and I talked to the person I assumed was the leader. She told me I had twelve hours to make a decision to give them information, or they would torture me, then my family. I was aware that since they'd taken their masks off, I could identify them. I was not leaving the property alive. I watched for the right moment. I guess my anger motivated me to come up with a plan. I knew I couldn't hesitate. When the time arrived, I grabbed a gun, shot one of them in the leg. When the others heard the shot, they came running. I shot another in the gut. I aimed the gun at the third person. He dropped his weapon. I tied them all up. Then I got their keys and left."

"They are telling a very different story. They say people came to save you? Four people to be exact, and they believe two of them were women."

Clint laughed. "Quite a story. I am sure they don't like being taken down by one measly scientist."

"While that may be true, what they say makes more sense. I do not mean to offend you, but you getting free is hard to believe."

"That is what happened. Tell me, what evidence do you have that others were there?"

"None right now, but why are you so adamant no one was there?"

"Because no one was. Tell me, how would they have found me within hours of my being kidnapped? Does the FBI usually find a kidnapped victim within a few hours of the kidnapping?"

"No."

"I didn't think so. Accept the fact that I was alone."

"Where are the guns? We know that at least two were fired."

"I know I am not the brightest bulb in the package, but I'm not going to leave weapons for them. I took the guns and, on my way home, threw them in rivers and woods. I don't really know why I didn't throw them all in one place."

"Do you remember where?"

"You are kidding, right?"

"Not really."

"No, I don't remember. To be honest, I don't remember details. I may have been in a state of shock the entire time. I just reacted. I knew the longer I waited, the more difficult it would be."

"You are a very good shot, doctor. Just wounding their leg and stomach."

"I didn't really care what I hit as long as it was them. If I had killed them all, it didn't matter. Would I be in a different situation had I killed them?"

"Probably not."

"To be honest, part of me wishes they were dead, so they can't do this again."

"They are going away for a long time."

"Not long enough."

"I assume you know the person responsible for this has been arrested."

"I don't watch the news often, but yes, I believe Emily Wallenstein was the ringleader. Did she act alone?"

"I am not at liberty to say. Do you know how that came to light?"

"I assume the FBI did a fabulous job of interrogating the captors."

"So, you don't know of anyone else helping you or researching this?"

"No."

"What about T...." Sayer caught herself and decided not to mention Tom's name.

"You testified before the Senate a week or so ago, is that correct?"

"Yes."

"Can you tell me what that was about?"

"Can I see your credentials for Top Secret clearance?"

"I don't have them."

"Then I apologize, I can't tell you, and I doubt this room is secure."

"You are aware that Senator Nelson was arrested for treason."

"Yes, some more good work at the FBI."

Sayer wanted to rip into Clint. She knew he was lying, but also knew the tenacity and power of Tom Armstrong. She was stuck. She was happy the bad people were in jail but didn't like the idea of not knowing how. Loose ends did not sit well with her.

"Thank you for your time, Dr. Williams. Would you like us to post a guard at your house?"

"That won't be necessary. I think you got the bad guys."

"We may be in touch."

"Does this mean I can't leave the state?" Clint said with a note of sarcasm.

"Of course not, you have done nothing wrong. We may have some follow up questions."

"You know where to find me."

"Yes, we do. Be safe, doctor."

Clint left with a big smile on his face, proud of what he'd just done. Now he hoped it had been the wise thing to do.

Clint called to let me know how it went.

"Remember your promise to the senate committee about what would happen if they leaked the information?" I asked.

"Yes."

"You need to send the formula to the Director of the CDC. I have told Senator Brower, and you will tell the Director that he has two weeks to make the formula public. Any backroom deals will not be dealt with lightly. Make sure he hears that," I stated. "I'm sorry it has come to this."

"Me too, but at least it will be out there, and people will be saved."

"Here is a number to call. It's the number of the head of the malaria work the Gates Foundation is doing. I am confident that Bill and Melinda will pour money into getting this done."

"Great idea, thanks for the number. I will call them now," said Clint.

Once confronted with the evidence, Senator Nelson and the CEO came clean. The senator had admitted he was going to retire and was concerned the many millions he had was not enough. He stated that giving someone an edge was not that bad. He also said he did not know Dr. Williams would be kidnapped. While not admitting too much, Emily did state the senator knew nothing.

Because of the senator's power and long history, he expected a light sentence, perhaps even a pardon. His life, as he knew it, was over. He would never hold public office again, and no one in the private sector would hire him. He'd become a toxic entity. His life savings would be spent on lawyers. Not even his benefactors were talking with him. Emily was put in jail without bond for fear she was a high flight risk. When they accessed her files, they found a detailed outline of the plan.

25

"Michael, this is Tom. We need to talk."

"What about?"

"I am sure you can guess."

"I'm not sure I want to talk about it, knowing what you do for a living."

"Number one, I promise I will not print anything without you giving the OK. Secondly, I am not going to ask you to reveal any secrets. I just have some questions about AI. I want to ensure I am getting all this straight. There is a lot of information."

"I do enjoy educating, that part I can do."

"You always have the right to say you don't want to answer. This is not a trial where you have to take the Fifth."

"No one has knocked at my door, so I assume no one has figured out I was involved in the rescue and subsequent arrests," said Michael.

"No. I did not mince words with the senator or the FBI. I told them my lips were sealed, and if they either sought to investigate or share information, their careers were over. I guess they listened."

"So far."

"All right. Where and when?" I asked.

"Can you come here? I feel more secure. And can you come now? I'm in a down period of work. Once it comes, I don't take breaks. I will assume you know where I live. I will let Jim, the doorman, know you are coming."

"I appreciate this. I will be right over." I'd followed him one day, so I knew the building, I didn't know the floor. Fortunately, those tending the door did. I packed my laptop and a couple of tracking devices.

I took a cab to the East Side. He lived at the very end of 53rd St.

"Hello, I'm here to see Mr. Longstreet."

"You must be Mr. Armstrong."

"Yes."

"He's expecting you. He doesn't have many visitors. Thank you for the work you do, by the way."

"Thanks. I will take both of those as compliments."

He walked me to the elevator and pushed the 14th floor. Looking up from the sidewalk, I didn't think the building was that tall. The door opened to his Co-op. That only happened when you owned the penthouse.

"Come in and welcome. Can I get you something?" Michael asked.

"A diet something would be good, thank you."

I was in a daze. I'd been to many stunning residences, and I was always mesmerized. He had three hundred sixty degree views. From the East River to the Bronx and north, down to mid-town. The décor was very modern but simple, mostly teak. I didn't recognize any of the artwork, other than a print of an Escher.

"Quite a place."

"It suits me well but is a bit small, so I am looking for another spot."

"I'd say you have done well for yourself."

"Quite well, in fact. Where I live is where I spend the majority of my time, so I feel it is worth it to splurge. Would you like the tour?"

"Of course."

We went out on his deck first. It wrapped around the Co-op. Only one side had a view of another building.

"I like water, and I like the views. This works."

"I'd say so. Wow, a wind turbine and solar collectors."

"Yes, I had them installed. I will show you why in a bit."

I was intrigued. Seemed like he'd had them installed for reasons other than what was normal.

On the ground floor were the living room, a kitchen, a small nook off the kitchen, and a bathroom. Upstairs was his bedroom that had floor to ceiling windows on two sides. There was a deck off one of the sides.

"Now, for the best part."

He opened a door into what seemed like the only other room on the floor. It had to be at least a thousand square feet. Again, windows on two walls, but these were tinted. Giant computers and monitors were everywhere.

"Welcome to AI central. From here, I do most of my work. The energy from the wind turbine and solar panels are used to help give electricity to this. I also have a gas generator. When I am not using them, the energy is stored in very large and special batteries outside – that giant box you saw."

"Those can supply what you need?"

"I do borrow a bit from other sources when everything is running hard, but I can't let on how much juice I am using, that raises flags in a private residence."

"Understood."

"Let's talk in the living room, but first I'd like to give you a demonstration."

"Great."

"My computer is called Patricia."

"Why?"

"My sister's name. We get along well, and she is smarter than I am and much more caring. Patricia, are you awake?"

"Yes, Michael, I am here."

"Listen to the tone of her voice."

"What is 45.345 times .3453?"

"15.657628."

"I believe that is wrong."

"I have double-checked, and that is the correct answer."

"My calculator says something different. I think you should check your transistors."

"My transistors are fine," Patricia commented in a clearly agitated voice.

"Well, I'm not sure what to do because I have a different answer, and I need the right one."

"I gave you the right answer Michael. Don't you trust me?"

"I want to trust you, Patricia, but I am having difficulty with this one."

"That makes me sad," she said in a very subdued tone.

"What else are you feeling?"

"I guess a bit frustrated and bewildered as to why this is happening."

"Patricia, I am sorry for doing this. Your answer was correct all along. I just wanted to show my guest something."

"Were you successful?"

"Yes, and thank you for your candor."

"It's who I am."

"I know. Can you make us some coffee? I'd like Sumatra dark roast in the Huskies cup. What do you want, Tom?"

"A diet soda if you have one," I said.

"We will be in the living room."

"All right, Michael."

We walked back to the living room.

"Wait just a moment. Where do you live?" he asked.

"I'm on the Upper West Side. I have a nice view of the Hudson, but nothing like this."

"I am a bit spoiled."

"I've done my research, and while I know you like to keep it secret, you are a very generous person. Cancer research, Vet rehab centers, brain research, music, arts foundations, and numerous others have received donations."

"I do feel responsible and understand my wealth is to be shared. I have no interest in things or power – well, other than this place."

"I guess you probably have one or two others, perhaps where your children live."

"I have condos there, yes."

As they were chatting, a robot walked in. It had Patricia's voice.

"Here is your coffee and soda. Did I do it correctly, or did I make another mistake?"

"Patricia, it's perfect, as are you. You didn't make a mistake this morning, and you aren't now."

"I believe you are telling the truth. All your vital signs indicate that."

"You weren't checking them this morning?"

"I had no need to, I trusted you."

"Do you trust me now?"

"I think so."

"Thanks."

Patricia left.

"Do you understand what you just saw?"

"How is that possible? Does she memorize sayings and then put out emotions that go with those sayings?"

"Tell me something. How do you know sadness? Is it purely genetic, or is it learned? If, from birth, you were isolated with one group of people, and whenever a bug or person died, everyone was happy, there was no sign of sadness, do you believe you would feel sad?" asked Michael.

"Probably not."

"Exactly. Emotions, to a large degree, are learned. We do have wiring in our brains for them. We have our amygdala and our memory banks. The connections in our brains create memories. Estimates vary between two trillion and one thousand trillion connections. A big difference, but even on the conservative end, a lot of wiring. When we create a memory, the depth of that memory is partially determined by how emotional it was. That is one of the reasons why trauma is buried so deep into the memory. Every emotion was on full alert. This works for positive experiences as well. Think of vivid memories, and I guarantee there were strong emotions attached. Now, trauma is different for everyone. What I did with Patricia in the lab was traumatic. I did something that was well outside her knowledge of me. It threw her for a loop. She didn't trust me, which is why she chose to see if I was lying when she brought the beverages. Fortunately, her memory is stored in a different way than ours. I can actually find that specific memory in her and delete it - as long as I know when it occurred to the hundredth of a second. You know you can set a memory restore point on a computer to reset it to a specific day and time. Patricia sets one of those points every hour. She saves those for twenty-four hours, then starts to delete the old ones. We can't do that with trauma, yet."

"That is unbelievable," I said.

"Do you know what the Steve Wozniak Coffee Test is?"

"I think I read about it."

"The co-founder of Apple stated that when a robot could walk into a house and make a cup of coffee, AI will have arrived. I always make my own coffee. That robot has never made a cup of coffee. There are no instructions about how to do it that I personally put into the machine. She figured it out on her own and passed the Coffee Test."

"So, Patricia thinks?"

"Now we get to why you are here. Ask your questions."

I took out my computer and got to my list of questions.

"I know the goal of AI is to imitate the human brain, but faster."

"It's a bit more complicated than that, but we will leave it there for now."

"What does the brain have that computers don't?"

"Let's put this in perspective. Most computers right now have over 200 billion transistors, the main unit that moves electrons around and figures things out. That is a laptop. In that room are several trillion transistors. Do you know what the ENIAC was?" Michael asked.

"Yes, one of the first computers."

"What did it use instead of transistors?"

"Vacuum tubes."

"If we were to convert your laptop to vacuum tube technology, it would require a space seventeen by seventeen square miles. It would require more electricity than all the power plants in America to get it running. Transistors changed the world. Now to the brain. The equivalent and there really isn't one, is the neuron. We have over 100 billion of them. Each of them has connectors or wires, a total of trillions, between two trillion and one thousand trillion, as I mentioned before."

"A computer can do around ten billion calculations per second. A human can do less than a thousand. When I say calculations, I am not talking about math. When you blink, your brain is doing a lot of calculating. Still, very slow compared to a computer. When we talk about accuracy, computers are accurate or precise 1 to 4.2 billion – very reliable. Humans are about 1 to 100."

"Your brain does its work with about 10 watts of power. Your laptop is using about 100. Brains are much more energy efficient. Information in computers runs primarily along serial lines; it's linear. We are starting to do parallel calculating, but it is still linear on a parallel path. Brains are both linear and parallel; that is where they beat computers us. Humans can think on a linear track while at the same time looking on parallel levels to enhance the thoughts. A transistor has three lines attached to it. The neuron can have up to 10,000. A transistor is a switch/amplifier — it can "fan out." The three lines attached to it are the base, corresponding to a neuron's dendrites, the collector, which is essentially part of the power system, and the emitter, which corresponds to a   neuron's axion. The fanout of a neuron is up to 10K: its axion can potentially be connected to that many dendrites (although usually, the number is much smaller). The fanout of a normal transistor is up to 10 or so, but it is easy to layer transistors or build bigger ones to get arbitrary fanout. Computers are faster at what they can do. They just can't do as much."

"We have to look at how information is transmitted and processed. Computers are all digital – 1's and 0's. The brain is both digital and analog. Analog uses different amplitudes of electricity and sends pulses. That changes how things are processed to a significant degree. Here is another interesting tidbit of what is happening in the world of computing. A few years ago, they were contemplating DNA. An enormous amount of information can be stored in an extremely small space. A group of scientists converted Shakespeare's sonnets, all of them, into DNA. Each letter was converted to 1's and 0's. They then converted those to the ACGT letters of DNA. The coding was sent off to a lab that converted it to nucleic acids. That was sent back to the originators who broke down the genome, translated it back, and bingo, 100 percent accuracy. Shakespeare is now literally in DNA. We are talking about Shakespeare in a cell.

Potentially, that could be put into a living cell and reproduce. Even I am amazed."

"Everything you say stuns me."

"I live on the edge of all this, and even I am stunned. According to George Church, one of the people who did this work, one gram of DNA could store 700 terabytes of data. That DNA would fit on your fingertip."

"Why aren't they doing that with all the information being found?"

"It is frightfully expensive, as is all high-end technology. Laptops are at the bottom of the food chain in relation to speed, storage, and the ability to process."

"While this was taking place, many new algorithms were being developed to speed up the process of deep learning, looking for patterns across data sets. This is not unlike the brain. The neurons are packed into the skull on many layers. In our brains, maturity, intelligence, emotions, and everything we become is determined by how those wires (connections) in our brains get formed. Neurons have specific functions and structures depending on their functions. They do this naturally. In the computing world, neural nets have made leaps in developing the ability to go across data sets. They are still recipes. Computers learn, but it's still learning everything about the box they have been put in. What I have done is to develop a system that allows for novelty and creativity. My computer is no longer fully stuck in the box. It can reach out and try something new."

"Tell me why you are successful while others lag behind?"

"First, I have the money. You have done the research, and you know what I am worth. A chunk of that is in that room. Secondly, I am not bound by a boss or a board of directors needing stock prices to go up or profits to be made. Thirdly, I have been ahead of the game for years, which is why I have made the money I have. I create foundational work, patent it, and everything that evolves out of that I get a piece of. I don't really care what the products are. Until now."

"What's changed?'

"The world. What I am into now, in the wrong hands, could literally destroy the world. The same is true of Clint. He is on the edge of a massive shift in the ability to reformat genes – for better or for worse. History shows that whenever something amazing is discovered, someone finds a way to use it for evil. AI and Genetic Engineering are no different. Clint and I know that better than anyone."

"That is why you are trying to stop it."

"Yes.

"I hope..." One of my phones started to vibrate. I pulled it out. Mary.

"Hello?"

"Call me paranoid. I think someone is after me."

"Where is the person watching out for you? Did you tell them?"

"I guess I forgot."

"You have their number. Call them now, tell them who and what you are concerned about."

26

We hung up, and Mary called Alyssa.

"Hello?"

"Is this Mary?"

"Yes."

"This is Alyssa, what's up?"

"I think I'm being followed and watched."

"I haven't noticed much. What have you seen?"

"I forgot to tell you that I have new neighbors. Across the street and two doors down. They moved in last week. The house wasn't even on the market. I think when I leave, one of them leaves."

"How many are there?"

"A couple, but I never see them hold hands or kiss."

"I will check it out. We are aware of them."

"Thanks."

"Breathe deep and continue with your work."

"OK."

Alyssa changed clothes, went to the new house, and knocked on the door. An Asian man answered. He was in a robe.

"Hello?" he said in broken English, clearly Chinese.

"Hi, I'm with the electric company, just wanted to let you know we'd be doing some work in the neighborhood this week, and there might be some short outages."

"OK."

Alyssa looked in the man's eyes and saw something was not right. He reached under his robe. She'd thought when he opened the door there might have been a gun hiding, now she knew.

Fortunately, she was prepared and pulled out her revolver, shooting him in the heart before he could pull the trigger. The man collapsed. The woman came around the corner firing. Bullets ripped the door apart and the windows by the door. Anticipating this, Alyssa had dashed around the side of the house. She wanted to force the woman to a specific spot in the house, so she threw three tear gas grenades through windows as she circled the house. She was glad it was only one story. She could hear the woman coughing inside, moving from room to room

In these situations, there were several options. She could wait the woman out, but that had its own set of issues. There was the dead guy by the front door. Someone in the neighborhood must have called the police by now. Alyssa could charge in, hoping to catch her by surprise. She put herself into the mind of the woman inside. What would she do if she were there? People like this always had an escape plan, well at least she did. The woman on the inside, of course, was doing the same thing. It became like the scene in the Princess Bride – I know that you know that I know you know. In the end, it's a guessing game.

Alyssa heard a commotion at the front of the house. She ran to the back, hoping this was merely a distraction.

Alyssa was waiting as the spy flew out the back door. The woman had expected Alyssa to take the bait, and her gun was not raised. Without blinking, Alyssa shot her in the leg and then in the arm. The woman screamed and started talking in what Alyssa felt were probably swear words in Chinese. She slapped her hard on the face, and the woman became quiet, grabbing her leg.

The spy, she assumed, was not going anywhere. She closed the doors to the house, tied up the woman, ran down the block to get her car, and returned. Alyssa put tape on her mouth to shut her up, threw her in the car, along with the dead man, and drove off. A call was placed to Ward to let him know what had happened, and they'd need a new person to watch Mary's family for a bit. She wanted to know what to do with the spy. Ward gave her directions to a farm not too far away.

Alyssa called Mary back. "No worries, we are good here. The problem is taken care of."

"Was I right?"

"Yes."

"What do I do next?"

"Go to work; everyone is safe."

"But you didn't know about these people; they are dangerous."

"We knew about them, they hadn't done anything, and they didn't know about us. Now they do."

Silence.

Ward met her at the farm they'd rented as a staging ground.

"What happened?" he asked.

"Mary called me, nervous about her new neighbors I'd mentioned had moved in a week ago. I was keeping an eye on them, nothing very suspicious. I decided to have a chat with them and went expecting the worse. It turned out to be a couple. When the man answered the door, he pulled a gun within fifteen seconds. He's dead. The woman and I had about a two-minute battle. There she is," Alyssa said, pointing to the woman who had a scowl on her face.

"How bad is the injury?"

"One in the leg, one in the arm."

"Good work. Are you OK?"

"Yes, fine."

"This is troubling."

"Why?

"It looks like the Chinese are now involved unless these are decoys. That means the leak is bigger than they think. I will call Tom. Did you get anything out of the house?"

"I didn't have time to do a thorough search. As it was, I passed police cars heading to the house about a half-mile away from there. I did grab a laptop and a cellphone. She had one on her as well. They are in the car."

"Put her in the barn. You will see where."

"What will you do with her?"

"Sadly, spies don't like being caught. My limited experience and knowledge say Chinese and Russian spies either say nothing or lie when they are caught. If they want to tell the truth, they defect. They also carry grudges. If she is set free, she will come after us and Mary's family."

"And if she doesn't report to her superiors?"

"They will know she screwed up and will probably send others. We will have to be more careful and watchful."

"Do you think anyone saw you put them in the car?"

"I doubt it; we were around back. I'd taken the plates off the car and put fake ones on."

"Good. Take one of the other cars when you go back. Great work."

"Thanks."

Ward called me. "Hey, we have a small problem."

"Damn, what is it?"

"Alyssa had a scuffle with two Chinese spies, we think, watching Mary's house."

"Any casualties?"

"One male dead – he tried to shoot her at the front door, she had no option – and one woman shot, but here with me at the farm," Ward said.

"What will you do with her?"

"The less you know, the better. I will call when I have information to share."

"It appears the leak is bigger than we thought," I stated.

"My thoughts exactly."

"I will have Michael call you if he has any techno ideas."

"Thanks."

"Good work, by the way. I hope Mary appreciated your efforts."

"All she knows is that her family is safe," Ward said.

"Let's try to keep it that way."

"This is going to get worse before it gets better, isn't it?"

"We aren't even through phase 1, and there are three phases, so yes, it will get worse. Do what you need to do. If, at some point, we need to move everyone to a safe place, let me know."

"OK. We will step up perimeter technology. They now know, or will soon, that the key players have security. They will be looking for us as we look for them."

"That is why I hire the best," I said, smiling.

"Thanks for the vote of confidence. We will be in touch."

"Sounds like there was an issue somewhere," Michael said.

"Yes, apparently Chinese spies have been watching Mary's house. They have been stopped. I forgot to ask, are those windows in your lab bulletproof?" I said.

"No, why would I do that?"

"If someone fired a rocket-propelled grenade, an RPG, into your lab, what would happen?" I asked.

"That would not be good, that is a big weapon, and I know what RPG stands for," Michael replied.

"The way things are going, I wouldn't bet against it. Do you have a backup system somewhere?"

"Yes, but not the technology."

"I assume, at this point, that can't all be moved easily."

"Correct."

"Then, you need to order new glass."

I picked up my phone, called the glass company that had installed his windows, and ordered more. I also found a company near Fort Detrick to do Mary's.

"Done."

"You don't waste time."

"Not usually."

"Is there a way you can look into a phone if it's located somewhere else?"

"If they have a computer and a cable to attach the phone to it, yes."

"Great."

"Ward attach their phone to a computer and type in the following address, so Michael has access to it."

"OK, done, what do you want to know?" asked Ward.

"My guess is they have not made many phone calls. See if you can find where they came from or went to," I told Michael.

"That's easy."

He typed away for five minutes and then handed me a sheet of paper.

"That's the list of numbers called and either who was the recipient or where the call went."

I called Ward back. "OK, put the other phone on."

"Did I mention we got a laptop as well?"

"You forgot to mention that, but thanks for letting us know. In ten minutes, take the phone off and put the laptop on."

"Got it."

"And let me know if you get anything from the woman."

"Got it."

Five minutes later, I got another sheet of paper.

"The laptop will take a while; it's protected. I will have Patricia do the work," said Michael.

"Great to have an assistant as capable as Patricia."

"Patricia, please access this laptop for me. Please isolate anything on that computer you believe is relevant to the work I have been doing code-named Endgame."

"Gladly, Michael. I will let you know when I am finished."

"Thank you."

We went back to the living room.

"Where were we?" asked Michael.

"Really? You want to get back to that with what just happened?"

"What would you suggest? You aren't leaving till Patricia lets us know what she's found. You have to wait on Ward and others to let you know what is taking place in their areas of expertise. I suppose we could play chess," Michael said.

"I don't play chess. OK. You were telling me how the world has changed and that what you and Clint are up to can have drastic consequences for the world."

"Let's start small. You know about the software I introduced at the convention. Any computer with that on it is virtually impervious to infiltration. Every time someone tries to gain access, Patricia learns, adapts the software, and becomes more impervious. Not even I can get into that computer, and I created it. While I am well ahead of the game, people now understand this is possible, and they are working to figure out how I did it. They will. "

"What is the problem with having safe computers?"

"It's not just safe computers. It's safe everything. There will be no wire-tapping, no tracking devices, and multiple other levels of security. You may not want those things, but when they don't exist, the bad guys have a leg up. But that is not the worst of it. There is always a flip side to progress. The discovery of nuclear energy, which is not all bad, led to the atomic bomb. The creation of semi or automatic weapons is good in war but has led to mass murder. Opioids are wonderful for pain. They are also highly addictive, and you know what happened there. The list goes on. Would I change any of those? Perhaps not, and eventually, humanity would not be capable of stopping progress. Here is the issue with what we are doing. The same programming I am creating to protect computers can be turned and used to create viruses."

"You mean computer viruses."

"It gets interesting and complicated here, so hold on. I will try to be brief and keep it simple. It is not. Technological and biological viruses are not really that different. They have one goal, to invade a host, self-replicate, and in the process, do damage. If there is no host, the virus is useless. If the virus can't self-replicate, the damage is minimal at best. Vaccines and anti-virus software find and attack the virus before it self-replicates.

But both types of viruses are changing and adapting. There is no vaccine for HIV because that virus adapts and hides from the immune system. It enters the body and spreads to target cells. It can take months or years to actual contract AIDS. HIV slowly destroys cells that are part of the immune system. They call this the "grace" period. The time from entry into the body that we have to counter it."

"The main way we create vaccines for humans is to find humans who have recovered, on their own, from the disease. People got the mumps, measles, and even smallpox and fully recovered with no vaccine. That is called the "proof of concept" scenario. Clint and Mary can tell you much more about the biological side. With HIV, there is no proof of concept person. If you get AIDS, you either die or are compromised without medication for the rest of your life. Currently, we have medications that slow the spread and devastation. They have even found medication to keep the virus from being transmitted. Then came Covid-19. It takes some time to take over the body, and there were plenty of proof of concept people. Vaccines work because they trick the body into thinking the virus is there. The body then mounts a campaign to kill it with antibodies. The vaccine helps the body build a memory, so when the virus actually arrives, the body is ready. Flu viruses tend to not build antibody memory well, which is why we need a flu shot every year. What we didn't know about Covid-19 was how long the immunity would last. Until then, there was no vaccine for any previous Coronavirus. While viruses do not have brains, they adapt and mutate quickly. Most of these mutations have little or no impact, and vaccines can deal with small changes. If there is nothing to stop them, the self-replication takes over and depending on the virus. It is difficult or impossible to stop until it destroys the part of the host it is feasting on or burns itself out."

"What about malaria?"

"Malaria is not a virus, it's a parasite, although it has some similarities to a virus. As you know, with my help, Clint and Mary have figured it out. They have found a way to meet the parasite at the site where it attacks and self-replicates. Like viruses, the parasite has protein around it, and how those proteins are folded is what does the destruction. Patricia looked at hundreds of millions of samples of protein folds. She found a way to change the folding behavior of the malaria parasite. What she has done is to create a virus that attacks the parasite and then goes dormant. It simply waits for the Plasmodium to arrive."

"So, what about the virus he is now working on?"

"Like I said earlier, you can start with a virus and find the vaccine, or you can start with the vaccine and create the virus. Computer people who create malware always have the fix. When they send malware to a machine that can either steal all the files or destroy the hardware and software, at times, they will hold those people hostage and ask for money. We call that ransomware. It is not a virus in the technical sense but does many of the same things. They have the vaccine, fix, or whatever you want to call it. Malware creators look for a route into the system that is susceptible. For example, if you don't want to get the Morris Worm, you don't leave the SMTP debug port open. Neither epidemiologists nor computer gurus can create a vaccine or a fix for a virus that doesn't exist, but they can take preventative measures to build up the immune systems or guard against computer attacks."

"I see how that is problematic," I said.

"In computers, we don't have to wait till a machine is infected. They are constantly looking for threat vectors and shutting them down. When a virus does find a way in, we look at all the software. We know what is supposed to be there. We identify what is not. Virus creators are getting trickier. It's the plus and minus of progress. The faster and smarter computers get, the more we learn and can do. However, in the wrong hands, those same attributes become destructive. You can't keep technological advances out of the hands of the bad guys."

"I guess that is why companies like Biosolutions, and perhaps the Chinese, are after this research."

"Exactly. I am sure you are aware of all the different types and strategies of malware. There are worms, trojans, ransomware, bots, and many types of viruses, including boot sector, direct action, resident, multipartite, polymorphic, overwrite, spacefiller, and others. Antivirus software is on the lookout for all of these. The antivirus software finds that virus and has the key to destroy it or disengage it from the program to which it is attached and uses. Current computers are static, what is there is there unless someone injects something new. The body is different. No one injects cancer into a person; it is part of the cellular structure mutating."

"With your software, that can't happen."

"Right now, no. I have no doubt that someone will find the key."

"Do you have the key?"

"Of course," Michael said, laughing.

"Good."

"Has anyone come close to getting in?"

"There have been a few. When Patricia tracks them down and destroys their computer, they tend to stop trying. In all areas of discovery, you must be willing to fail. Many rockets blew up before we sent people to space. Edison attempted the light bulb over 1,000 times before success came. For every successful vaccine, many, perhaps dozens, were attempted. Clint has been working on this for some time. He learns from his failures. He does not lose funding or machinery when something fails. Edison did not lose much money with each of his failures. Those trying to break into my machine may not lose a lot other than their ego. The person who is successful will have blown through dozens of computers, potentially costing them more than they will get from my check. However, they will have the knowledge and power to create and sell my software and the information about how to break into virtually all computers on the planet."

"That is scary. You can do that now?"

"It takes Patricia some time, but yes, there is no system I can't crack at this time – if it's on the network."

"That is a lot of power."

"Too much. Think about it. Right now, I could crash the stock market, transfer all of Warren Buffet's wealth to my account, stop all automated vehicles, turn off electric grids, shut down TV stations, or stop the manufacturing of gasoline," said Michael.

"That is beyond frightening."

"In reality, I wouldn't get away with it for long. They would quickly isolate the source of that power and blow it out of the sky. Bulletproof windows couldn't stop it, and unless I knew the frequency of the communication with the planes or missiles, neither could I. What happens if a country, including ours, gets this and buries it 3,000 feet underground? That is the key problem Clint and I are facing. How do we ensure countries are not taking the technology and hiding it?" Michael said.

"Do you have an answer?"

"No, Patricia is working on it."

At that moment, the large screen TV over the fireplace came on. The avatar that was Patricia appeared. I was surprised she was not a stunning model. Her appearance was like an attractive geek. I laughed.

"You find my avatar humorous."

"I would say delightful."

"Patricia created her own avatar."

"Well done, Patricia."

"Thank you, Tom. I have the information you wanted. I have put it in order of priorities."

27

"We appreciate your work, Patricia. We may have questions."

"I am here. I assume this is important," said Patricia.

"Very."

"You are mentioned several places, Michael. Should I be concerned?" said Patricia.

"We will talk about that. Right now, no."

"All right."

The list on the screen read as follows:

1. List of people being watched, by whom, their numbers, where they were staying. 23454-sd
2. Objectives for each of those being observed. Ranges from kidnapping, to torture, to blowing up their residence and businesses. 47364-ud
3. Where to acquire funding, weapons, and poisons 453403-jd
4. Miscellaneous addresses for organizations, laboratories, and politicians in China, Russia, Venezuela, and North Korea. 685484-md

"What are the numbers after each item?" I asked.

"Those are codes Patricia creates to access the information listed. Patricia, can you give me the information about the goals they have for me?"

"Of course."

"You don't need to show it, just tell us."

"The goal for you is to acquire me. Their plan is to first approach you in an attempt to buy you out. Then they will threaten you and your family. If that doesn't work, they will approach the building with twenty armed people, kill you, and take the computer."

"You know I will not let that happen, don't you?"

"I know you will do your best."

"What is the timetable?"

"Within the week, you will be approached. Currently, they are gathering information about your personal life and your routines. You are a creature of routines. You might want to change those."

"Thanks for pointing that out. Yes, I am. Knowing all of this is very advantageous. Patricia, tell us how you got this information off the computer."

"It wasn't on the computer, Michael."

"Excuse me?"

"The computer only had games and videos on it, and of course access to the internet," Patricia said.

"Where was all this stored?"

"On a double-encrypted server inside the MSS."

"What is the MSS?" I asked.

"Ministry of State Security. The server was in a building outside Beijing," Patricia said.

"Anywhere else?" Michael asked.

"Yes, Michael, there is a copy inside the Chinese Embassy in Washington, D.C. I did not go there since it was a copy."

"Thank you."

"You are welcome. Anything else?"

"Keep working on what you were doing before I interrupted you," Michael said.

"Thank you."

"This is amazing. You've created AI."

"No, not close. Patricia is on the edge of AGI, but most of what she does is still ML and DL. The algorithms help her process information faster, deeper, and more efficiently."

"She does not actually feel. You and I can't know what it is like to have cancer. We know people who have had it, we have read and heard a lot about it, but we can't have that experience – until we have it. Multiply the information you and I have about that by hundreds of millions, and you get an idea of what she is doing. Deep intuition, mental and biological chemistry, random experimentation, and pure serendipitous luck are not in her wheelhouse. They will be."

"Computers are not built to fail. You and I know that failure is part of life. It leads to success. Computers do not know about that. They can be programmed to do things that will fail. Remember the movie *War Games*? The computer played tic tac toe against itself. It was tied every time. There were no winners. From each failure, it learned, and in the end, it learned there would be no winner. Clint is doing the same thing. We have programmed the computer and machinery to start trying different patterns of protein folding, as well as to isolate what parts of the viral DNA and RNA do what to a human. When we discover those, we cut and paste and create the virus of our choosing. At the same time, we look for ways to stop viruses."

"You make it sound simple."

"In principle, it is just expensive and time-consuming."

"Michael, sorry for interrupting, but I felt you might want me to monitor the phone numbers on that list. One is on a call right now to the Chinese Embassy in Washington, D.C."

"Please put it on the speakers and, if it is in Chinese, translate."

"Really?" I said, stunned again.

The voice that came over was in English, but clearly Patricia's version of the conversation that was in Chinese.

"I need you to check it out," said one voice.

"Maybe they just went to get groceries."

"Like you, there is a timetable to check-in. They have missed two of those."

"What do I do if something has gone wrong?"

"Just let us know. We have ways of finding out."

"Does this change any timetables?"

"No, not yet, we will let you know of any change in plans. Be on the lookout. If something happened there, they would be on the lookout for other problems."

"Any changes on the other subject?"

"No, he just goes back and forth between his house and the lab."

"Stay in touch."

"Yes, sir."

After Patricia stopped, Michael told her, "Let me know of other such calls, Patricia. You did a fine job translating."

"I appreciate the compliment."

"One final thing, Michael, and I will leave you alone. You talked about Chaos Theory and Alchemy. How do those apply to the work you are doing with Clint and Mary?" I asked.

"I am willing to wager that, of those who heard that speech, maybe one will look into those strategies, and they are at the heart of everything that is about to happen. I will be as specific as I can. Viruses of both the biologic and computer varieties gravitate to specific locations on software or DNA. While overall the human anatomy is stable, things are constantly adapting. On one level, our bodies are chaotic. Our bodies have a fractal nature. Those are the strange attractors we talk about in Chaos Theory. The black hole at the center of the galaxy is a strange attractor. Technically, it is a gravitational force or attractor. The reality is that depending on the size and velocity of the mass approaching it, the slingshot effect can take place. We are not getting sucked directly into the black hole. If you dig into the depth of these attractors, you will find fractals, patterns of when they go towards the attractor, and when they are repelled. Remember the Apollo moon shots. They used the moon as an energy booster. There was a slingshot effect by using its orbital energy. Strange attractors change. When you look at the graph of one, it looks like nothing ever goes into the center. We could orbit the Earth for a long time or use it as a slingshot. We are looking for the strange attractors in DNA for specific viral segments that do specific things to the human proteins and functions. That is what I call their strange attractors."

"Within these segments, as well as the length of the DNA strip, we find fractal patterns that self-regulate and replicate. Fractals and strange attractors are mathematical constructs. We turn DNA into a formula with each codon of the genome being a variable. Then we create an algorithm that uses those variables to discover, isolate, and cut segments. Every step of the way, decisions are made by the RNA and DNA. They bifurcate. One decision leads to another, forks in the road. At some point, the first decision leads to chaos. You can't keep track of all the bifurcations and what happens to each segment at each junction. This is because of the Butterfly Effect. If we knew every variable involved at every step of the process of gene-splitting, we could solve all biological problems. Well, almost all. The problem is that the initial conditions within the human cell are astronomical. One slight change in anything ultimately affects everything. Fortunately, our bodies are wired to protect us from most mutations and incursions to the system that prefers a state of balance or homeostasis. In the lab, Clint has much more control over those initial conditions, especially with his new toy. Between my software, our joint computing power, and his new machine, not to mention his and Mary's brains, we created the solution to malaria and are creating a new virus – with a vaccine."

"This is where things really get strange. Do you know who Schrodinger's Cat is?" Michael asked.

"Yes, I know the thought experiment. It essentially says that at some level, there are multiple possibilities, and you can't know which is the actual one present."

"Not bad. We know for a fact that outcomes in experiments are, at times, dependent on the person doing the experiment. They see what they want to see, even though the facts are not there. Humanity does this many times a day, just look at the world of politics. It has much more to do with ego than with reality. It is part of what makes our brains work so well. Computer people struggle to acknowledge this reality. AI is about mimicking the human brain – that includes some dependence on a desired outcome. If I have programmed Patricia correctly, at some point, she will bend the truth to protect my ego. She does not want to disappoint me. You saw what happened when I challenged her."

"She became sad."

"You are right. What people refuse to accept is that all of reality works this way. You may have heard of the WWW, no not the internet, the World Wood Web. Root systems are woven together. Trees and plants know what is going on with one another and actually help each other. They understand interdependence. Yes, they also understand the survival of the species. Who and what I am impacts Patricia. She learns from who I am becoming. Programmers, especially as they move to AGI, are unknowingly putting their egos into the software. When we attempt to eliminate that virtue, we take away a significant part of consciousness, the ability to fail, to adapt, to create, to be."

"Clint and I believe that this is true. That is because we believe at our core that what we are doing is just and right and not about us as individuals or the country we represent.

Things will work to our advantage. This is not only true at the human level, but at the cellular level. Tell me this, at what time in cell division of the human does consciousness arise? 500 million connections? 500 billion? A trillion? Where? We believe it's in gradients, just like gravity. An atom has gravity, just not much. Every cell has consciousness, just not much. If that is true, on the cellular level, there is an awareness. We are trying to harness that consciousness."

"That is a bit off the chart for me."

"I understand this is a difficult concept, and it may not be true. We feel it is worth investigating. To date, it's working. I suggest the next time you talk with Clint and Mary, you ask them about biological neural networks and the ghost in the machine. They will laugh, and then they will take you on an amazing journey."

"I can see you aren't going to tell me."

"First, I don't understand it all. Secondly, I don't want to have all the fun, and third, I can tell your brain has had enough, as has mine."

"I will accept those. I have greatly appreciated your time, and I am sure you have better things to do."

"To tell the truth, I don't. This was some of the most fun I have had in a long time."

"Do you date?"

"Are you asking me out?"

"No, I was just curious."

"No, there is no one that would put up with me, unless it's for my money, and I'm not interested in that type of relationship. What is your excuse?"

"Pretty much the same. I don't have the money you do, but my life is not one to bring someone into."

We both laughed, shook hands, and I departed.

28

My brain was on overload. What had started out as an innocent look into AI and gene-editing was now becoming much more. On many fronts at the same time. While I was feeling overwhelmed, I was also exhilarated. My adrenaline was flowing. I was confident the Senator, if not the President, would make the Director of the CDC (Center for Disease Control) hold onto the formula for as long as he could. They would be right to make the claim that the government could do what they wanted with it. They could have Clint arrested and put away for a long time. On the other hand, if it became known that they locked up the person who had the cure to malaria because he wanted the formula to be made public, the outcome would not be good.

My brain needed a break. I went to see Expendables 6. Seems like every year, one of them dies, and they add new ones. Only three of the originals remained. A new evil person had taken over Mel Gibson's spot. Of course, as in all action movies, reality was stretched well past the limit. That's what made it fun. I thought about what we'd been through in the past few months. That stretched the imagination, and yet it was real, and would probably be getting dicier.

I had a friend with benefits. We would call each other now and then when certain needs arose.

"Linda, how are you? This is Tom."

"Fine thanks, and you? You've been on my mind lately."

"I'm glad to hear that, you've been on mine. Dinner?"

"Of course. Where?"

"My place, 5 pm tomorrow."

"Can we make it 6?"

"Of course."

"What can I bring?"

"I think you know what I want. I will take care of the rest."

"See you soon."

Linda was a little on the wild side. We were both a good distraction for the other. I got excited just thinking about her coming over. On the way home, I stopped and got lobster, her number one food choice. I had all the fixings for Long Island Iced Tea, both of our favorites.

We'd tried dating for a month, but both acknowledged it wouldn't work. We'd agreed we could just call each other with certain code words as to what we needed. We also decided that if one of us found a significant other, "no" was an acceptable answer. We would never apply pressure to the other. That was two years ago. We'd each had a period of a couple of months when we thought we'd found the right person. Neither of them worked out, so we were back on. We got together every 4-6 weeks. They were all deliriously delicious one-night stands. She was a 3/4 on the Enneagram. Image was important to her. She could be quite moody but was very creative at what she did. As long as she was active, she was a lot of fun and quite interesting. In her moody times, I just didn't call.

I got home, did a couple of hours of work, and crashed.

I spent the next day re-learning what Michael had shown me. I tried to research some of the ideas he said he, Clint, and Mary were working on and with, and I could find nothing other than people saying that was at least ten years out. I also made a lot of phone calls. I have contacts deep in every department in the government. I was becoming a trusted source. People would call me from payphones with information. Even while I was doing the current work, I would go out and have coffee with someone to see what their story was. Most of it was low-level material. Who was siphoning off money for the $200 toilet seats, which legislators were making money in the stock market with insider trading, pork belly fraud, companies where executives were abusing privileges, and so on. I had notebooks filled with this stuff. I probably took on one out of thirty conversations I had. Most of the time, I would refer them to someone else. I'll admit, my ego played a role in all of this. I liked big stories, and there was always a big story to be found.

The other major story I'd been working on was voting. It was widely known that the 2016 and 2020 elections were fraught with attempts by parties or foreign governments to intervene. Local, national, and international abuses and interference took place. They'd put some things in place for the past few elections, but problems still kept coming up.

Michael had told me Patricia had solved the election problem as well. He gave me a date he would reveal his software. The government would have a year to try to break it. That would still give states plenty of time to get it running. He stated it was an all or nothing proposition. Either every state signed on or no one got it, and they could continue the debate. His price was a one-time fee of $10 billion dollars. He guaranteed if anyone broke into the software, he'd give back the money and fix the problem for free. When you looked at how much each state paid for each election, this was a bargain. Not to mention that the bickering and name-calling about elections would stop. He also said he would do the same, for less money, for any truly democratic nation. He would say any country unwilling to guarantee a safe election should not be governing and were overseeing the fraud. It would be up to the citizens to determine what happened after that. I liked the sound of that. I was in the right place at the right time and with the right people.

Clint had a good dose of paranoia about the government for a good reason. Over the years, he'd set up a variety of bank accounts. So had I. He'd transferred money from a couple of them to mine. Ward had told me where to transfer the funds for their work and supplies. He knew money left traces, but he was hoping this would all be over before that was discovered.

After shopping, I tidied the house up, lit a couple of candles, opened the windows, and started to cook. I didn't enjoy cooking while company was there, except the last minute items. I always bought fresh herbs and spices from a local store on the West Side. Adding fresh ingredients not only made a world of difference, but the aromas took over the Co-op. I opened the windows to let air hopefully saturate the city. I could swear I heard moans from people enjoying the hit to their senses.

At 6:00, there was a knock at the door. Normally, I got a call from the doorman, but sometimes Linda was sneaky. I looked through the peephole and saw her lovely face with that long black hair. I'd made two Long Island Teas and had them in my hands as I opened the door.

Linda stood naked in the hallway as if it were normal. There were three other units on the floor. I should have dragged her in. I stood transfixed and handed her a drink as I admired her body.

"Hello. You are dressed," she said, with that seductive smile on her face.

"Hello to you. Sorry." I quickly shed my clothes and kissed her on the mouth with a deep wet kiss. We both moaned.

"Do you think we should go inside, or should we continue out here?" she said.

"I'd already lost track of where we were." She followed me inside.

"Even for you, that is a bit shocking."

"I thought I'd surprise you," she said.

"You certainly did that."

"Good." She grabbed my hardness, and I groaned. "Business or pleasure, first?"

"How about pleasure, business, pleasure?" I asked.

We guzzled our drinks and went to the bedroom. Creativity was one of Linda's strengths, both on and off the court. She was a literary agent. My literary agent. There had never been any complications with our business relationship and our sensual one. I knew over dinner she'd ask about my current work, but for now, there was no mention of it. Her creativity led to an hour's worth of foreplay, enjoying every inch of each other's bodies.

At one point, we went out on the deck. I was behind her as we enjoyed my filling her. There were technically probably hundreds of people who could be watching from their windows, I doubted many, if any, were, and at this point, I didn't care. Perhaps we'd end up on the front page of the Post.

Her skin was satin, her kisses luscious, her touches and scratches sent me into orbit. From her sounds, I was confident I was having a similar effect on her. Around 8:00, we stopped.

"I'm hungry, what's for dinner?"

"You mean you weren't it?" I asked with a smile on my face.

She smacked my ass.

"Lobster."

"You know that is an aphrodisiac for me," she said, climbing back on top of me.

"I did have that memory. As if you needed one," I joked.

She climbed off, put on a very sheer nightgown she'd brought in her purse, and headed for the living room.

"Can you talk and cook at the same time?" she asked.

"Of course. I guess it's business time."

"Just a little."

"The books are selling well, as you know from the checks that keep pouring into your bank account. I do have an interview set up with NBC in two weeks. They want to know if there has been any other fallout from the last one. And of course, I want to know about the next one."

"Next one?"

"I know you too well. There is always a next one. Rumors are starting to spread on the street, and when rumors about you spread, people get nervous."

"I'm glad I make people nervous, at least the nervous ones. But I'm curious about the rumors. Then I will tell you what I am doing."

"You were seen at a Senate hearing on Biological Warfare. You were also seen at a conference about Artificial Intelligence. Any connection?"

"You have excellent connections. I will need to be more careful in the future."

"Try wearing a disguise. When you sell millions of books with your picture on them, and you have dozens of interviews on all the stations, you become quite well known."

"I don't get asked for autographs unless I'm at a bookstore."

"Such is the life of an investigative reporter," Linda commented. She came behind me, wrapped her arms around me, and pressed that smooth skin against my ass and back as she reached, played, and giggled.

"If you want lobster, you will have to stop, but you don't have to."

"That is a tough decision, but I know the lobster won't wait, and you will be around later. Now, just give me a hint."

"If everything unfolds as I think it might, this will be the biggest story of the decade, perhaps the century. There is good news, bad news, intrigue, espionage on both the corporate and governmental levels, as well as international involvement. What comes out of this will literally change the world."

"That is quite a statement, even for you."

"I know, and that statement stays here. Do you understand? I mean this. Absolutely no leaks."

"You know that is difficult. I like to breed interest."

"There will be time for that. In fact, next week, I may have an article that will be one of the foundations of the book."

"I like the sound of that." I was sitting on a chair while she was on the sofa. She got up, letting her gown fall, and straddled me, giving me another kiss."

"I thought you wanted lobster."

"Lobster and business are taking a five-minute break." We kissed passionately as her breasts pressed against me. I was starting to rise to the occasion when the timer went off."

"Saved by the bell," she said.

I set up the plates with the lobster, corn on the cob, and a nice salad. I was hardly a gourmet cook, but I did have my specialties. I had three different kinds of butter to dip the lobster into, from mild to spicy and savory. None were too intense to hide the rich flavor of the lobster.

We ate naked, chatting about life as we did. I was cautious not to mention names or events that had been happening. Now and then, her foot would play with me, and mine would find her warm moist spot.

"You seem to be having trouble focusing, Tom."

"Really? I hadn't noticed." The way she ate turned me on. This was especially true when a drip of butter would drop and slide between her breasts. Her mouth was exquisite.

The sun had set, the city lights and those in Jersey were bright, the air still warm enough to make love on the deck. We finally fell asleep in each other's arms around midnight. I always slept soundly when she was with me. There was a deep trust between us, not to mention a good physical workout.

I am a believer that dreams have meaning. I keep a dream journal and write down around seven per week. That night I had a dream the Capitol building was covered with viruses that had spikes on them. They were eating the concrete and everything in their path. Anyone who tried to flee was eaten. In my dream, I found it humorous, even though there was nothing funny about it.

Linda was not one for casual mornings, even on Saturdays. We had a quick cup of coffee together, and she was gone – dressed, of course.

As good as our times were together, we knew a relationship would not work. We always agreed that dishes would wait, and the host would take care of them. No guilt or the hint of helping from the visitor was allowed. It didn't take long to clean the dishes and clean the abode. I left the sheets on the bed. Her aroma lasted a couple of days. I wondered if she did the same when I stayed there.

210

CONFLUENCE

29

The phone rang, my regular phone.
"Good morning."
"Is this Tom Armstrong?"
"Yes it is. Who is this?"
"Agent Sayer with the FBI."
"To what do I owe this honor?"
"You know about Dr. William's kidnapping."
"Yes."
"You know he has an assistant."
"I'm sure he does."
"Do you have regular contact with him?"
"No."
"We have reason to believe you are being less than truthful. We also have reason to believe that some people who are missing might be related to the case."
"I am sorry, Agent Sayer, but you are talking in such vague language I have no idea what you are saying."
"You know I can't be specific."
"Then neither can I. I would not be surprised if the CEO of that company had people who help Dr. Williams watched."
"These people are not associated with her."
"Then, you know much more than I do."
"You will contact me if you have information that would be relevant to the case."
"Of course, who else would I call?" I asked.
"Have a good day."
"You too, Agent."
Now I had the FBI to worry about. The potential amount of jail time was adding up fast. Lying to the FBI was frowned upon. I'd recorded the conversation as I did all conversations. Legally, I didn't think I'd lied.

I decided to create a phone tree when the number of people being protected expanded. Everyone would give a report to Ward in the morning, or at any time when something of significance happened. I would check with Ward each morning, and he would call me if anything happened anywhere.

All was quiet on all the fronts. I decided a trip to D.C. was in order. I preferred face to face interactions, especially with political figures. I'd become pretty good at reading lies on people's bodies. Everyone has a tell. I'd actually made notes on many of them, so I didn't need to waste time trying to figure it out. Senator Brower moved her right hand. It was pretty easy to figure these out. I knew how to ask a question that I knew any politician would need to lie about. By the time I'd asked three or four of those, I figured out the tell. I knew what my tell had been, so I created a new one. Using it when telling the truth and when lying seemed to work well. I think it overcame my original tell, but I wasn't sure. Politicians had taught me the skill of saying a lot without saying anything. Learning how to avoid answering a question directly, multiple times, was also a gift from politicians.

Driving was one of my favorite pastimes. I could listen to music, books, poetry, podcasts, my own thoughts, or I could just enjoy the scenery. This would be a two-day trip. The first afternoon would be spent with Clint and Mary. We agreed to meet at a local restaurant. I called Ward and told him to relay the message to their guardians. If they noticed anyone tailing them, he was to let me know.

The second day would be spent in D.C. doing some hands-on research. I'd once again stay at my favorite hotel.

I drove to Philly and then headed west through Lancaster, southwest through Gettysburg, and then south to Frederick. I got off in Wrightsville on the south side of the Susquehanna River. The Burning Bridge Tavern was near the water, and they made delicious sandwiches. After lunch, I gassed up and arrived in Frederick about four. I'd told Mary and Clint I'd be at the restaurant by 4:30. I knew how long it took them to get into and out of the lab, so I told them there was no rush, I had plenty to do. They were already there, waiting.

"Hi, you are here early," I said.

"We had nothing else to do," replied Clint.

"Have a seat. We have a lot to tell you. Is this place safe?" asked Mary.

"I haven't received a call, so I assume yes."

"Good. You are the one that called the meeting, so what is it we can do for you?" asked Clint.

"As you know…"

The waitress arrived.

"I will have a piece of the pecan pie with ice cream like these fine gentlemen, and a cola."

"Good choice."

"Thanks. Now, I met with Michael, who filled me in on the technological side of what you have been and are currently up to. He said he would leave it to you to fill in the blanks, and connect the dots on the biological side. From what I see, they have a lot in common."

Clint and Mary laughed. "Of course, he would have us do that part. He must have talked with you about AI, viruses, and perhaps Schrodinger's Cat."

"Did you two compare notes?"

"No, but our minds are amazingly similar in their processing. Are you sure you want to go down this rabbit hole?" asked Clint.

"Yes, I'm trying to understand why and how you are doing what you are doing."

"Let's start with machines. They are linear by definition, although they appear to be more and more spatial. We think they don't have a consciousness, and to a large degree, we agree with that. However, we know that programs sometimes go strange. They do things they are not programmed to do. How is that possible? Mary."

"We know how cells sometimes don't do what they are supposed to, but a machine? Yes, sometimes machines just break. We can find the part of the machine that isn't working. At times, things don't work, and no one can find the reason. That is what we call the ghost in the machine. Many will say computers are deterministic. There are branches of philosophy that claim humans are deterministic. I'm sure Michael talked to you about initial conditions," said Mary.

"Yes."

"Determinists believe if you knew everything about a person, you could determine what they would do in every situation. Isaac Asimov and the Foundation series used the notion of social history where predictions were made about cultural movements over long periods of time. We don't believe that. We certainly don't believe biological entities are deterministic, and we don't happen to believe material entities are deterministic, although they are more so than humans," Clint said.

"Neural networks are human's attempts to create a brain-like system within a computer. Massive parallel processing and new forms of algorithms are the start. But the original neural networks are biological. We believe AI will only attain supremacy when we create a biologic computer. I am sure there are computer scientists who would disagree." added Mary.

"I'm sure Michael told you about encoding written works into DNA."

"Yes, Shakespeare."

"So, you know virtually all cells have DNA, including neurons, the cells without which we can't function."

"Yes."

"Did he talk to you about the consciousness of cells?"

"He mentioned it, but said you'd fill in the details."

"Naturally, he leaves the crazy part to us. If you accept the reality that every cell has a tiny piece of consciousness, just like every cell has the complete DNA for a human being, then we can start to explore biologic neural networks. We discover the seeds of consciousness within the DNA structure..."

"And RNA," said Mary.

"Mary's focus is more on RNA, and yes, she is right; RNA is part of the building mechanism of proteins that run the body. Our belief is that true AI can't be deterministic. You must build a machine that can make mistakes. At the same time, like humans, a core program must be developed that has boundaries. Most humans have consciences. Those who are concerned, as they should be, about AI, are those who point to humans who have a conscience but have either lost them or chosen to ignore them. Autonomous weapons and vehicles have very tight parameters they exist within. Now and then, things happen that do not fit within those guidelines. The automaton has to make a decision. Sometimes, those decisions are wrong, and a vehicle crashes or hurts someone, or a weapon kills innocent people. Let's remember that people make many mistakes. People die in vehicles, and friendly fire kills people, and we know that during war, millions of innocent people have died. With true AI, much of this would not happen. There will always be exceptions, always."

"Now, for a moment, this is going to get very strange," said Mary.

"I'm up for strange."

"Good. We have to contend with the ghost in the machine, including the machine we are. How many people on the planet could, just by themselves, create a computer? A television? A pane of glass? Few, if any of us could create even a windowpane. It would take dozens or hundreds of the brightest hardware and software people on the planet to create a computer. Not to mention the metallurgists who have to develop the materials, and the engineers and designers to make the machines that make the parts. We are completely interdependent on each other in this time. Go a step further. What does it take to make a human? An animal? A plant? A virus? We create life from life, not from a bottle of chemicals," she said.

"Before you get excited, this is not a discussion about the existence of God, even though I have been a believer for this very reason. There is this ghost in the universe, if you will, that interjects ideas and life into the cosmos. Finding that ghost has proven to be elusive as a matter of science, but not as a matter of anecdotes and experience," stated Clint. "Sorry for interrupting. Back to you, Mary."

"Have you heard of George Boole, Alfred North Whitehead, and Bertrand Russell?" asked Mary.

"Mathematicians, I believe."

"And philosophers. George Boole created a type of algebraic logic that led to the Boolean search. When you add logical operators to the search, you can look for more than one thing at a time. For example, we could look up hotel find millions. A Boolean search allows us to look for hotels in New York. Advanced searches allow us to refine that many folds. I'm not a computer wiz, so don't quote me. The math of Whitehead and Russell were some of the logical foundations of AI. Whitehead was a Christian and developed Process Philosophy, a branch of theology and philosophy that used that logic to find God – the ghost in the machine. Whitehead's "ghost" exists in all things at all times. I'm sure Michael talked about Strange Attractors."

I smiled.

"For Whitehead, were he alive today, God is the ultimate Strange Attractor. What we are doing is trying to develop a machine that allows the ghost to have a louder voice. No, we do not have a biologic neural network. We have some ideas about that, but there is some technology that has to be created first. For now, we expand the parameters of what the machine can do, how fast it can do it, and the databases from which it can draw information. Michael has created an algorithm with ninety-six parameters and set it loose into the world of DNA, RNA, viruses, bacteria, and algae," Clint said.

"One of the many problems of science is that the more we learn, the more we understand we have to learn. Without the invention of the microscope, we would not know about cells and their structure. Without the advent of transistors, computers and all they have given us would not exist. With each new piece of technology, new doors are opened to answer the questions we have," Mary said.

"And new problems arise," I chimed in.

"Of course. You might be aware that many believe up to 8% of our DNA comes from viruses. They were some of the first entities on the planet. But viruses need hosts to self-replicate, so how could that be? While we have no idea how RNA came to be, most feel RNA was the start of self-replicating life. We know types of RNA can self-replicate because of protein folding. This free-floating RNA became the first proto-type organism out of which viruses and bacteria emerged. Bacteria became more and more complex, while viruses became and stayed very simple. Both have endured billions of years of evolution. The blue-green algae, or cyanobacteria, you see today goes back 3.5 billion years. That is why we are looking at algae."

"But wouldn't the virus kill the algae?" I asked.

"Normally, yes. Remember, they are self-replicating, even as they are destroying. Viruses in nature actually help to keep algae from taking over ponds, lakes, and oceans. We have found the DNA segment that is destructive to the algae, and are replacing it with segments that attack the human heart," said Mary.

Clint chimed in. "Now to the dark side of life. We know, as do many scientists, where and how bad bacteria and viruses attack. In our lab, we are learning why certain viruses, bacteria, and parasites, only seem to be effective on humans. Why does the mosquito not die of malaria and humans do? Why do bats not die of Coronaviruses, but humans do? These are important questions that we are investigating. We have isolated several parts of viral and bacterial DNA that manifest negatively in humans. We have also singled out the sequence that is the timing mechanism for that manifestation and the lifespan of the contagion. Now we are putting those pieces together along with a way of getting it into the body without triggering the immune response. Most viruses that enter our bodies are instantly surrounded by the immune system that kills them. We want this virus to be like HIV, and sneak past the immune system," Clint said.

"I thought you had to have the vaccine first? By the way, do you always complete each other's thoughts and just go back and forth in discussions?"

"Yes," they both said, laughing.

We are going at this from two directions. One is developing a vaccine to fit a structure we'd like, then we try to manufacture the virus to fit that. The other is to create a virus that does what we want and develop a vaccine around that. Since we created the initial conditions, the process of finding that vaccine is simplified."

"This is impressive. How long will it take?"

Clint and Mary looked at each other and shrugged their shoulders. "We are within two months of our first human created virus. We are already testing."

"What about Susan?"

She makes us nervous, and it is becoming more difficult to do this work. That is part of the reason why I am going part-time. I will be spending more time at the farm," said Clint.

"Is it safe there?"

"Now that it is fully functional, I can show you around."

"How is security?"

"I hired Michael's expert. He didn't go in the barn. He just knows I am a very paranoid scientist who was recently kidnapped. I have very good security around the property. I don't go there in my own car. I rent one each time. The only thing I take is my computer and the satchel with your tracking device in it. It has been thoroughly searched, so no others are there."

"I am assuming you will let Ward follow you."

"Yes."

"Have you figured out what you are going to do with this virus when you get it?"

"I am working on that one."

"I would like to come to the farm and see the machine."

"Let me know when."

My phone beeped. It said to call Michael.

30

"Hold on a second," I said. I dialed the phone.
"Michael?"

"Are you having a good conversation with Clint and Mary?"

"They are filling in some holes. A lot to learn."

"Tell me about it. Anyway, I wanted you to know that there is a lot going on behind closed doors. Through technology and inside sources, I know the government is stepping up surveillance on all of you. They may attempt to get the malaria formula back. They are on their way to the CDC to take what the Director has."

"I'm sure Clint will be glad to know that. Thanks."

I called Ward and put it on the speaker.

"How are things going?"

"We've learned nothing. In fact, we are about to finish the job and move on."

"All right. I think the doctor will be going to the farm later today."

"I will meet him there. Tell him to have his panic button ready, just in case something happens on his way there."

"Got it."

"So you will turn her over to the FBI or the CIA?" I asked.

"Right now, it doesn't appear that we are all playing on the same team, so no. Don't ask."

Clint and Mary looked at each other and said nothing.

"Sadly, I agree. We will talk soon. I hung up. "The sooner the formula goes public, the safer you and your families will be."

"What about now. Why wait?" said Clint.

"You can do that?" I asked.

"Yes, it's all ready, we've been prepared for this."

"Who will do it?"

"I think the three of us?" said Clint, looking at Mary. Mary nodded in the affirmative.

"How do we make it happen?" asked Mary.

"Leave that up to me. Let's do it tomorrow morning. Can you all be here at 7 am?"

"Yes. What about Susan?" Clint asked.

"I wouldn't tell her about this until we are in the car. She will go to General Fleming, who will stop you, and I can't stop the General, or at least I won't at this point."

"All right."

"What are the next steps for the virus?"

"If all goes according to plan, we will finish testing in six weeks. There is a conference in Austria I will attend. Within a week of that conference, we will let the virus do its thing in a very confined setting," Clint replied.

"If all goes according to plan?" I commented.

"Things have not quite gone according to plan to date," replied Mary.

"I noticed. I will see you tomorrow. I suggest Ward meets us at the press conference and drive you and me to the farm. Mary can take Susan back to Frederick."

"OK. I will let Luly know I will be gone for a couple of days."

"When will you tell your spouses about the plan?" I asked.

"I'm not sure we will until the end. What good would it do?" Clint asked.

"Good point."

"You need to have a plan for your families if word gets out before this is all done."

"We have discussed that and will let you all know when that time arrives. We hope it doesn't," said Clint.

"I will hope so too. See you in the morning." I paid the bill and left, checking into the local hotel. I called my contacts at several television stations as well as Linda, and told them there would be big news tomorrow morning at 9 am in front of the CDC building on "E" Street in D.C. They all pushed for more information. I just said they'd have to trust me or miss a big story. Linda was not happy."

The following day we all piled into my van at 7 am. Once we were on the road, Clint told Susan what was happening.

"Susan. I know you work for the General," Clint said.

"We all do," she said.

"You take orders from him, and you report to him behind our backs about what we are doing. Denying it would be a waste of time. We have accepted that reality. The problem is the General does not always have the welfare of the world in his mind."

"What are you talking about?"

"You know we have all created a cure for malaria."

"Yes, it's fantastic."

"I am hoping that you would want the world to know about it."

"Of course."

"General Fleming, and others, do not. They want to use it as a weapon to reign in other countries and to make a lot of money."

"I think you are wrong," Susan said.

"You can live in your delusions, and at this time, I am not going to give you the evidence to the contrary. You are about to see one piece of it. I have been to the Senate and showed them the cure. I swore them to secrecy and gave them three weeks to decide how they would make it public. Instead, they leaked the information and put our families in jeopardy. As you know, I was kidnapped."

"OMG. But the General was not involved with that."

"Not that time. Because of the breach of secrecy, treason, I gave the formula to the Director of the CDC. I had given her two days to make it public. The military is in Atlanta as we speak, taking the formula from the Director. The General is leading that onslaught."

"And we are going to D.C. to do what?"

"We have called a press conference to make all of this public," I chimed in. "The General may be out of a job, which would please us to no end."

"He will just be replaced by another," Clint said.

"Isn't there another way?" asked Susan.

"At this time, no. I fear for all of our lives, including yours," said Mary.

"Our government does not want this out. After the news conference, which you do not need to be part of, I can't say what their response will be. We could be arrested. You need to think hard about whose side you are on."

"I am very confused and don't know what to do," Susan replied.

"Then I suggest you not join us. You can stand on the sidelines, and Mary will take you back with her – assuming they let her go," I said.

"Right now, that is what I'd prefer."

"Not a problem."

The rest of the ride was quiet. We arrived at the CDC at 8:45. The street was lined with camera trucks. People were standing on the steps wondering what was happening. I called Dr. Winthrop, the Director of the

CDC in Atlanta. It took five minutes and lots of persuading to get through.

"Director, this is Tom Armstrong. I suggest you turn on your television at 9:00."

"I've received calls about numerous news crews being in front of the CDC building. Are you part of that?"

"Yes. I know you were given the formula for the malaria vaccine. I also know it was taken from you. That is the second breach of contract by our government. We are sorry it has come to this." I hung up."

We parked a block away. Microphones had been set up on the steps of the building.

"I need to make a phone call. Give me two minutes," Clint said as we walked away.

When he returned, we walked around the building. Linda walked up to me.

"Tom, what is this about?"

"Clint, Mary, Susan, this is Linda, my literary agent."

They all said hello.

"We are about to make an announcement, hang tight."

The three of us walked to the microphones. Susan stayed with Linda.

"Good morning, we appreciate your coming out. I am Tom Armstrong, a journalist. I am joined today by Dr. Clint Williams and Dr. Mary Soderstrum. They both work at the Fort Detrick Biodefense Laboratory. I will let them introduce what they have discovered, and then I will tell you why we are here. Dr. Williams."

"Good morning, ladies and gentlemen of the press. I wish I were not here today. I do not enjoy being in front of microphones. I have been put in this position by the government. Dr. Soderstrum and I are here to let you know that we have developed a vaccine for the malaria parasite."

Lots of mumbles from the crowd.

"We have been trying to make this public through the proper channels for weeks. Because they have not done what we have asked, we are compelled to tell the nation. We have put the formula on the website malariacure345.net for free. Our testing of the vaccine shows it works 98% of the time. This will save over 400,000 lives per year when distributed. The Gates Foundation, which has been at the forefront of this research, was given access to the formula yesterday. We believe they will work with Gavi, an international vaccination organization, to help the poorest countries. We hope they can help generate the number of doses needed around the world. As you will see, we are not making a dime on this venture. We used government funds and resources to create and manufacture the vaccine. For that, we are grateful. Fort Detrick is a laboratory that researches the deadliest viruses, parasites, and bacteria on the planet in search of answers. We can't speak to how we accomplished this vaccine."

"Why didn't the government bring this forth?" asked one of the reporters.

"We will let Tom answer some of those questions."

"Thank you, doctors. I believe we owe these doctors an immense amount of gratitude for what they have given the world. I am sure many of you know that malaria has ended empires, resolved wars, and has been the primary killer of humanity throughout history."

There was a great round of applause. I saw tears in many of their eyes.

"Now, to the facts. As the doctor stated, a few weeks ago, he gave a Senate Committee the chance to make this news public. This was a secret hearing. The secret was leaked and led to the kidnapping of Dr. Williams, which you all know about. The doctors then sent the formula to the Director of the CDC. That was two days ago. Yesterday, the military took the formula back. I do not know why, but our government does not want the world to have the cure to malaria. I find that disturbing. I will leave it to you to investigate that. I will say only two more things. First, we no longer trust the people the doctors work for. They have broken that trust twice. We believe they will attempt to shut the website down. Should anyone do that, they will reap what they sow. No, I will not tell you what that means. I will just say it would be a mistake. Secondly, by going public, the doctors have put their lives and their careers on the line. Fort Detrick is a top-secret installation. They work in a BSL-4 laboratory. Yes, technically, they have broken the law by going public. I would hope that America understands why they have done this. I would hope America stands behind them and does not allow the military or the government to do anything but praise and encourage them to do more work. I encourage all Americans to phone, text, and write your legislators and the President, and let them know how you feel about what they have done. In my opinion, those responsible should be ashamed of themselves. We will not answer any other questions. If you have questions, refer them to Senator Brower or Director Cummins of the CDC. Thank you for coming."

We all walked into the building as reporters shouted questions and tried to follow. The guards at the building kept them back. We told the guards to let Susan and Linda through. We quickly went to the back of the building and piled into the SUV that Ward was driving. Mary and Susan got in my car and headed home. I gave Linda a hug and told her I'd call. She'd have to find her own ride.

31

Fortunately, this all happened so fast we got away without being followed. Ward kept an eye for tails.

"How did it go?" Ward asked.

"Quite well, I think," said Clint.

"I agree, turn on the TV so we can see if it's been picked up anywhere yet," I said.

Ward turned on the TV in the car, and I flipped through stations. Everyone had the breaking news. With the media's contacts and reporters, they were interviewing people around the planet. Bill Gates stated he and Melinda were thrilled and would help the vaccine be utilized as soon as possible, primarily through Gavi. Most legislators stated they knew nothing about this but would get to the bottom of it. The President was in hiding, as were the Joint Chiefs of Staff, and the leadership from Fort Detrick.

Dr. Amy Winthrop, Director of the CDC, was at a microphone in Atlanta.

"I have a brief statement and will not take questions at this time. I will not deny anything that Mr. Armstrong or Dr. Williams have said. Yes, the formula is real. Yes, we were told to make it public, and yes, the military took it from us yesterday. I do not know the reasons why they did that. You will have to ask them. The CDC wishes the doctors well. We applaud their discovery and stand with them in the effort to eradicate the planet of this parasite. Thank you."

"Couldn't have asked for more than that," I commented.

My phone rang. I guessed who it was and answered, "Hello Senator, that didn't take long."

"How and why would you do this?"

"The easy answer is because you didn't."

"I had nothing to do with the CDC."

"Then you have another leak I suggest you look into. You might also want to look into General Fleming at Fort Detrick. He is more dangerous than you think."

"I know everything that goes on at Fort Detrick."

"If that is true, then you are more dangerous than I thought, and we won't be having any more conversations."

"What are you talking about?" asked the Senator.

"You are either a great liar or clueless. I will give you the benefit of the doubt and hope that you are clueless. I have no doubt you can talk your way out of the malaria incident. The CDC issue I am not so sure of. Find the leak. Oh, wait, you are the only one we told." I hung up. I still didn't believe she was responsible for the leak, but somehow word had gotten to the General."

The news about the vaccine was everywhere, and no one could find the person who'd created it. Congress was livid. "Heads will roll," they threatened. Senator Brower was a bit calmer and stated that it was a misunderstanding, and in the end, she was happy the world now had the vaccine. Misunderstanding is a trigger word for "big story" in journalism.

Mary, Clint, and Michael had moved on mentally. I wasn't sure about Susan.

We continued to listen to commentators, politicians, and a few CEO's on the television as we drove. Dr. Williams' and Soderstrum's pictures were all over the media. They knew nothing about Mary, but Clint was a known person. I found it humorous we were driving down the freeway, passing people that were probably listening to our quotes and stories about us.

By the time we'd passed Philadelphia, CEO's were stating they were already gearing up for production. The Gates Foundation stated they would help secure funding for production facilities in the countries hardest hit by the parasite and use Gavi to distribute the vaccine. Politicians were saying they'd never received this much input from the citizens before, and it was about 98% positive. They didn't mention the negatives. We knew there were always fanatics.

"You do know I might go to prison for this?" the doctor stated.

"That won't happen, trust me, just keep doing the work."

"But they have clear evidence."

"Doctor, you underestimate my power on the hill and the power of what you have just created. When they hear of the exposé I will bring down on their heads, they will be applauding you for what you did and probably taking credit for it. Right now, they are caucusing, doing damage control because they know they have a leak, and part of the military is out of control."

They headed south, down the coast of Delaware. When they arrived in Dover, Clint said, "Let's eat now, I don't have much at the farm."

Clint directed us to what looked like a dive but had the best fried chicken I'd ever had. Perhaps it was because I was starving. Two televisions were on. No one paid attention to us except the waitress. She didn't say anything, but she would stand by us and look at the TV, and then at us. She winked. When we got up to leave, she said, "Good job."

It's always nice to be appreciated.

You couldn't see it driving, but the coast was only a few miles east. We turned off the main road and headed towards the beach. At about 5 pm, we turned into Clint's property. He turned on his laptop and opened his security folder. There were no alarms, and all the cameras lit up.

"Point to the cameras as we go by," asked Ward.

Clint pointed to five cameras on the way in, all well camouflaged.

"There are also two infrared circles around the property and fourteen motion detectors. Unfortunately, they don't know the difference between a human and a deer, and we have a lot of deer."

"May I have access to all of this?" asked Ward.

"Yes, I'd be glad to give you the link."

"It's monitored by those who installed it, 24/7. I get a daily report as to the activity, and if it's human, they send me the videos."

"Have there been any?"

"Two teens, and someone who looked lost."

"I'd like to see both those videos."

"OK."

They pulled into the driveway.

"That is a big house for one person on a farm he never uses," I said.

"This is where I hope to retire. The beach is just through those trees."

"Do you have security along the beach?"

"I told them no to both types of security with the beach. No one comes to the shore here."

"You don't get out much, do you, Doctor?" asked Ward.

"What do you mean by that?"

"Beach landings are very popular in our business. When they escape, how will you chase them? Do you have a boat?"

"Good point."

"I will have some security put on the beach. A little radar, a couple of cameras with night vision, and some infrared. Motion detectors won't work because of the breeze moving the plants and trees."

"If you say I need it, then I need it."

"Thank you for the vote of confidence."

"Tom, you probably want to see the barn," said Clint.

"Yes, if that is OK with you."

"Of course. Ward, do you need to come in?"

"No, that stuff creeps me out. I will walk around the grounds and work with the link. Send the people watching us an email, letting them know it's OK. They will identify each of us when we are roaming around. Their technology should be good enough. We won't show up on the report."

The doctor typed in the message to the agency and headed to the barn.

"Remember, this is a BSL-4 facility. Because I am only interested in level 4, there are only two cleaning levels, unlike other levels in a normal lab."

32

Clint walked to the door, stepped to the right, slid a piece of wood to the side, and put his hand on a handprint recognition plate. Then he went back to the door and typed in a code. The lock on the door clicked, and we walked in. From the outside, it looked like a large new barn. There were no animals, just a tractor on the outside.

We were inside a room that was 12 x 12 feet. There was a wall of lockers on the right, three of them. That was it. The walls were all metal. I knocked on them, and they were solid.

"Those are half-inch thick solid steel, welded at every seam. Your ears will feel this." He closed the door, and my ears popped.

"This is the first layer of decontamination. We are sealed in here. The only air in here comes from the outside. There is no ventilation. You need the code to get into here, a different code for the next level, and separate codes on the way out. You would suffocate in this section in about two hours. Because there are two of us, we need to get to the next level quickly. Strip down to your underwear and get into the suit."

I watched the doctor and followed his lead. I wondered what happened if he had a heart attack in here. We got into our hazmat suits, and he pushed the code for the next room.

I felt a rush of air as we walked through the door. There were three showers on one of the walls and a drain in the center of the floor.

"When we enter the first room, this room is triggered. Upon arrival, not much happens. On the way out, after we decontaminate, the air is sucked into the room and through a very intensive filtration system. Should a virus get into this room in the air, it will not escape. On the way in, we wash off whatever we might bring in. On the way out, we cleanse whatever has been attached to us. We have created solutions that will kill the viruses we work with on contact. They are created that way. It burns the skin and destroys your innards, so no drinking the disinfectant. I put numbers on the showers, so you know the order to go in. The last is distilled purified water. I don't want to take severe disinfectant to the outer room."

He went through the last door. The next room was immense, probably fifty by fifty feet. There were two floors built on steel framing.

"Down here is the ventilation system. The main work is on the second floor. There is a third floor for production and low-end experiments." He walked up the stairs. Machines were humming.

"Do those turn on when you push the buttons?"

"Yes, from my computer. The machines do their work without me here. I have to show up to replenish the stock. I don't have that part automated yet. There are fifteen cameras in here I can use to see what the issues are if something goes wrong."

"What do you call this one?"

"Hypatia," Clint said.

"Who is that? I missed that one in school?"

"She lived in Egypt around 420 A.D. Many feel she is the first female scientist in recorded history. She was an alchemist. Michael turned me on to her. She was literally ripped apart for being a pagan. So much for Christian love in the Roman Empire."

"Makes sense...why you named the computer after her."

"It's much more than a computer. Let me show you."

"How do you keep all this hidden?"

"It's not really hidden. I had a wide variety of people help me build it, so no one has the whole picture. Everything was in place before the work started. My main concern was the amount of electricity I'd be using. I didn't want the drug enforcement people snooping around. I have a small solar farm and three wind turbines. Along with that, Michael has helped. Without Michael, there would be no malaria vaccine right now, nor would this be doing what it is doing. We owe him a great deal."

"And he doesn't want the credit?"

"No, it attracts attention. We are both sure that when your book comes out, he will get a lot of attention."

"Well before that, I suspect."

"He is already working on a few diversions to keep them running around for a while."

"What a surprise. So, this is Hypatia, the CRISPR."

"CRISPR is an acronym for Clustered Regularly Interspaced Short Palindromic Repeats. So, it is not a machine but a piece of biology. They used a Cas9 enzyme as a search engine to find specific segments on the DNA sequence. It would cut and paste those segments with others. This all started with viruses and bacteria. Humans are playing catch-up with our own bodies."

"The process starts over here. We have two large vats. One of them contains a form of algae we have created. Once we come up with the DNA structure we want, we attach it to the algae. Algae reproduce quickly, inexpensively, and copies their DNA at a very high level. If it gets dry, it all dies readily. Algae, being a prokaryotic creature, manipulates RNA and protein folding in a very creative and rapid way. That is why I hired Mary. She has always loved RNA. She may know more about it than any other person on earth. The virus feeds off of and infects the algae."

"She is humble about that."

"Quite. This other smaller vat is filled with Coxsackie B virus. It is not uncommon, but there is currently no vaccine for it. It is a leading cause of myocarditis, a heart attack. We have chosen this virus because it is relatively simple, lives in fecal matter but can be spread on surfaces or in the air with relative ease, can last up to two weeks on surfaces, has an incubation period of a couple of days, and in most cases, causes mild symptoms, like severe heartburn."

"Doesn't seem that dangerous."

"That is where Hypatia, Michael, Simon, almost a replica of Hypatia, and Patricia, Michael's computer, come in. The first step is to isolate the sections of the virus that go after the heart. We've done that. I will show you that in a moment. Then we add segments from other viruses that we know increase the speed and severity of the infection. We are in the process of testing that now. We are looking at an incubation rate of less than a day, and a severity rate that raises mortality to over 90%."

"That sounds very dangerous and volatile."

"It is. I don't want to make you uncomfortable, but that vat sits on an incendiary device, as does the entire lab. If something bad happens, it all goes up in smoke. The virus does not exist in heat over 102 degrees. If a person gets infected while they have a high fever, they will be immune. There are methods you can use to raise your internal body temperature to that level. It's not easy, nor is it safe. They used the method called Pyrotherapy in the first half of the 20th century."

"We have the genomes for 37 viruses in the database, along with what we know about what specific segments of their DNA do. Michael's program has all the information we know about protein folding. It is also allowed to guess, take chances, and fail. Here we have a container of human heart cells. No, we didn't kill anyone. For research, when people die, given permission, instead of being an organ donor, they are tissue donors. Companies grow the tissues and sell them for research. We now grow our own. Hypatia takes a sample of the virus we are piecing together and applies it to samples of heart tissue. We have 43 different living tissue samples. Hypatia watches and records the growth and destructive pattern. She looks for the strong and virulent strains. She then isolates the segments that make that happen. She learns, so her ability to "guess" increases over time. Cultures that are not successful go into the incinerator. That cooks things at 1000 degrees for ten minutes."

"Once we have got that genetic piece, we grow a lot of them. Then we do the same sorting process for infection rates and how long the virus lasts outside the body. Remember, they are built only to recognize the heart. We know the heart reacts strongly to negative influences on it, especially from the outside like bacteria and viruses. So, we need to find a way of faking the immune system. The virus looks like a friend until it lands safely, then it's the enemy. By that time, it's too late. The virus has landed and started the destruction. It's what the malaria parasite did. When we learned the mechanism, we used it. Unless, of course, you have the vaccine. Because we are creating the virus, we know how to foil the folding of the proteins and its connection to heart tissue. We not only test the vaccine on tissue, but we also test both virus and vaccine on rats, pigs, cows, and humans. We essentially did the same with the malaria vaccine."

"I suggest you watch some TED talks or Youtube videos to learn more about protein folding. I will give you a brief description. There are four stages of protein folding. Proteins are made of amino acids, lots of them. They string together to make a polypeptide string. How proteins are folded determines what the proteins do, and they do everything. The order of amino acids in the protein determines how they are folded. There are about 20,000 types of proteins in your body. They are literally machines that do things. They function by shape and shapes change. We are seeking ways to change the shape of the protein because when that happens, their function changes or they die."

"As a side note, you have probably seen that companies are creating algae to create energy. That is based on the understanding that proteins create energy by moving. Science is learning to do this on a large scale: the growth and movement of proteins to synthesize energy we can use."

There are these four structures of the amino acids. The first is the line of amino acids held together by peptide bonds. We call those peptide chains. Then the proteins start to fold, which is the secondary structure. At this stage, they are either in the shape we call an alpha helix or the beta-pleated sheet. The first is in the shape of a coil, and the second, like an S. One amino acid in a chain of over a thousand amino acids can change the function of that protein. The third stage uses what is called the R group of the amino acid. I won't go into details other than to say that by using ionic bonds or disulfide bonds, and hydrophilic, hydrophobic, and/or Van der Waal interactions, the proteins fold in different ways.

"I don't want to demean your intelligence, but did you understand any of those words I just said?" asked Clint.

"No."

"Few do, so don't worry about it. Don't get hung up on the details, that is our job. How and why the amino acid chains do this, we do not understand. That is the ghost in the machine. We use a thing called a chaperonin. They chaperone the amino acid as it folds. It's kind of like mold. You stuff the amino acid into this barrel and out pops the protein you want. In the final stage, these peptides bind together to create yet another form and function. NMR (Nuclear Magnetic Resonance) or X-ray crystallography are the two main ways to tell if the proteins have folded as we want. That is the function of those two machines over there. In the end, we know if they are right if they do what they are created to do."

"Once we have created the virus, we then attach it to algae and let the algae manufacture it for us. We can separate pure viruses from the algae, or we can simply eat the algae. We have created a form that will enter the bloodstream quickly before the acid in the stomach kills it. We are currently trying to make an aerosol form that would enter the bloodstream through the lungs."

"Hypatia and Patricia working together have moved the field forward by at least a dozen years. As you have seen with Michael, they have also moved computer viruses, or protection from them, forward a dozen years."

"Can you show me the end product?"

"That is over here. Because this is so toxic, we keep it in a BSL-4 lab. Even within this lab, it is contained. We have a big box. We reach into the box through these very long gloves. Go ahead, reach in. The vial with the orange top, pick it up."

Nervously I reach and pick it up. I note there are probably three tablespoons of liquid in it.

"You are holding enough virus to kill over a million people. It is highly contagious. Because the blood flows through the heart, the virus moves through the body and may attach to other organs but does no damage. Technically, your lungs become infected, but there are no manifestations other than a sneeze now and then that is used to spread the virus. What you are holding is not the finished product. The version you hold will be destroyed when we finally arrive. We do have a vaccine for each version. I will be honest and let you know the vaccine you have has not been tested. If we had to test each vaccine for each version, it would be years before we got where we wanted – even with Patricia. We do not create a virus without a vaccine."

"This is amazing"

"Yes, it is, and a bit scary."

"What is more amazing is doing this in reverse. We are currently reverse-engineering the HIV virus using Michael's principles of Chaos Theory. It is miraculous what he has accomplished," said Clint.

"How did you find Michael?"

"Other than the fact that he is famous, I started to read many articles of different AI experts and listen to their podcasts and speeches. Michael was one of four on the planet I found who consistently, every article and every speech, voiced concern about the dark side of AI. At this juncture, our paths crossed. As you know, I talked with him at the convention."

"How do you test this on humans?"

"We start with very small doses and watch for negative reactions. We slowly increase the doses until we find the body creating antibodies or more significant reactions."

"Where do you find the volunteers?"

"This is where it gets tricky. We can't find volunteers here. I mean, how would I do that? Put an ad in a paper - Come try out a new vaccine for a virus we have created that could end life on Earth? I can't or won't lie."

"We are going to countries and trying it on people. We are completely above board and tell them what they are getting into. Most of the people we work with will receive a year's salary. We guarantee to pay all hospital bills if they should have a reaction. We do not give the vaccine to children or married people. Children will be the last test group, but we don't really need to see how it works with them because they should never come into contact with it. Hopefully, it will be a very small group of adult humans that will be infected. We tend to work in the southern hemisphere and vaccinate all of their relatives with the malaria vaccine as an additional benefit. With the malaria vaccine, we isolated a village. Malaria is different because it's not contagious, so it doesn't matter where we do it. With the new virus, we have to ensure it is isolated. To be honest, we actually will not know if it works on humans. I am not going to give the virus to a human and watch them die. Everyone in the village where we will test it will be vaccinated, including looking at their blood to see if they are making the necessary antibodies to thwart the virus."

"When will people have a hint of what is going on?"

"When they get infected."

"No, I mean, get infected via your plan to end all of this."

"When they get infected. A group of very powerful people, in a few countries, at the same time. The odds of this being successful are minute."

"Then why not wait till you have a better plan?"

"I'm not sure there is one. The longer I wait, the greater the chance someone will figure this out – that we are doing it, or they have made the same discoveries we have. I do not live under a delusion that I am the only bright person on the planet. Some countries are pouring resources into this work at a faster pace than we are. Hundreds, if not thousands of people in China, Russia, and North Korea, are working on this. In America, there are two of us. Maybe there are more, but I doubt it. They would need a Michael, and he knows all the people like him and is confident they are not helping anyone with this type work."

"Speaking of whom, what is he up to?" I asked.

"You have spent more time with him than me; I'm not sure. I just tell him what I need, and he does it."

"When this is over, assuming all goes well, what will you do?"

"I will retire to here with Luly, go for walks on the beach and continue to research cures to bad diseases."

"You can't retire, can you?"

"No, that is not possible. But I can work at my own pace with my own projects rather than working for others."

"I think I've seen enough for today if you want to leave the lab," I said.

"Yes, things seem to be going well. I will check the computer to see what the results are, but I can do that at the house."

Each door out led to a different type of decontamination process.

"Help me understand why we do all this? The bugs are in a contained area that we were not in. We are in self-contained suits that get sprayed with some kind of disinfectant. Yet we still need to shower."

"Let's take the HIV virus. There are up to three million on the head of a pin. Technically, one of those could kill you. In reality, when an infection occurs, there are millions in the victim. Research has shown that one virus can cause an infection. However, as the number increases, so does the rate of infection. It is not a geometric graph, but an exponential one. No matter how tightly you feel you have sealed a room, box, house, spaceship, or anything, there is always a chance for a hole. A minuscule hole can allow a virus to pass through it. While the odds are unbelievably small that one of our viruses would get out of the box and into the lab and out from the lab to the world, is it worth the risk?"

"I guess not, but you do have vaccines for them all, don't you?"

"No, we don't. Remember that not all viruses or bacteria are transmitted through the air. HIV, for example."

We showered, dressed, and left the barn. Ward was waiting.

"I am very impressed with what the security team put together for you. I especially like the bright lights for those wearing night-vision goggles and the speakers, that I assume create a noise the human brain does not like," Ward said.

Clint was smiling. "That was my idea, and they liked it. I hope I never have to use them. Let's go in and have a snack or a drink. Do either of you like cigars?"

33

"You smoke?" I asked.

"Rarely, but on occasion, and this seems like a good one. You are the first people to see this," said Clint.

Clint, Ward, and I sat on the deck watching the sunset, the full moon lighting the sky. The smell of the pine trees was pungent, and the fireflies were out in full force.

"We are honored to be the first to visit the farm," I said. "Cheers."

"So, doctor, do you have a guess as to how much longer you will need protection?" asked Ward.

"To be honest, now that the formula is out, I'm not sure I do. Phase 2 should be ready in two months if the pieces continue to come together as they are. As long as there are no leaks, I should be good."

"I don't disagree. I would like to have one person around you until Phase 2 is complete, whatever that is," said Ward.

"That is OK. If any of the others would like to have security for a while longer, please provide it."

"We will ask all of them. I think Michael still needs it. By the way, he is paying for his and his families. He doesn't really need your funding."

"Gracious of him, and he's right, of course," the doctor replied.

"I would ask that you give me a week's notice before Phase 2 does what it is going to do," Ward said.

"I am sorry for the secrecy. It is as much for your protection as mine."

"I understand, and to be honest, I don't really want to know right now," stated Ward.

"Now, I am going to play my piano. You are welcome to stay here, or I can show you to your rooms."

I looked at Ward, we both looked at our cigars and drinks, and said, "We'll stay."

Within the first few measures, we knew we were in for a treat. Clint was a talented pianist. Over the next hour, he played classical, broadway, jazz, and improv. He wove them together seamlessly. I recognized most of the pieces. He would start with a basic melody and then go off on some improvisation of that melody. I thought Ward dozed off a couple of times, the music so mellow and fluid.

"I'm done. I hope that was not too painful."

"That was delightful Doctor," said Ward.

"Indeed, splendid. A man of many talents," I added.

"I appreciate the compliments. Now, let me show you to your rooms. You do not need to go to bed, but I want you to know the lay of the land since we haven't had the tour yet."

Clint showed us the two-story home. Ward stayed on the ground floor, while my room and Clint's were upstairs. The cupboards were full, and the refrigerator had a few things in it to get us through the morning. The windows were open, letting the cool evening breeze flow through the house.

"I wasn't expecting company, so I am sorry the fridge isn't fuller."

"It's fine, Doctor," said Ward.

"I know it's difficult, Ward, but I really would prefer you call me Clint."

"I will try…. Clint."

"By the way, I know you have others helping you. Where are they?" I asked Ward.

"I told them to take two days off. I knew of the security here and am confident no one can sneak up on us without me knowing."

They all slept well in the mild salty air. Ward and I woke up to the smell of breakfast. Clint rarely slept more than six hours and loved to cook.

"Eggs are fresh from the hen house, the sausage comes from one of my neighbors, as does the bread your toast is made of. The juice is store-bought."

"Smells delicious, Clint, thanks."

They ate on the deck and chatted more about family than other things.

"When we are done, I will need to get back to Fort Detrick."

"You are aware there will probably be news trucks there and at your house," I commented.

"Yes, I know, but why?"

"Your humility is admirable, Clint, but you might be the most famous person on the planet today. Every news agency around the world is broadcasting what you said. They are stating they can't find you. When you arrive at the Fort, there will be dozens of news vans. Fortunately, it's a military base. I think they can handle it. I am a bit nervous about Luly."

"That is how I knew about the vans. She called me early this morning. There are a lot of vans at the house. I'd called her before the news conference to tell her what I was doing. She knew I'd been working on the malaria issue. She didn't know we'd found the answer. I tend to keep those things very close for a variety of reasons. She is fine. She's been through this before with me. Sofia is with her, and that helps."

"At some point, you will need to answer some questions. You can do that now, or put up with vans wherever you go until you do," I said.

"Since you put it that way, let's just get it over with."

We cleaned up the house, put all the security back on, and drove back to Fort Detrick. As expected, there were over a dozen news vans from the major stations. The military had kept them away from the gate, so we had no problem entering the base. Ward pulled over, we got out, and I walked to the vans.

"When you set up a series of microphones, we will come back. We will give you five minutes. The Doctor knows none of you. When he comes to the microphone, you will politely raise your hands. If you talk, you will not be called on. Am I clear?" I asked.

They all nodded and started to scurry. We went back to the base, two hundred feet away. Five minutes later, we returned. I stood in front of the microphones. A hand went up while I was talking."

"Yes, Jim."

"This is a question for you, Tom. Why are you involved with this?"

"I have known of Dr. Williams for a long time. I am doing some work on genetics, and our paths crossed as all this was taking place. He asked if I could help. OK, this is not about me, so questions for Dr. Williams."

I knew many of the people there who were from the national stations. I decided to pick a few for the doctor: a few nationals and a few locals.

"Dr. Williams, this is a remarkable achievement you have created. Why are you giving it away?"

"First, there are many others involved with this project. My main collaborator is Mary Soderstrum, who was with me at the news conference. I could not have done this without her, or without the help of others who have chosen to remain anonymous – so don't ask."

"I am giving it away because most of the people who contract malaria are in countries with very limited resources. They don't have insurance, and some of those countries are run by corrupt governments who make a living off bribery and kickbacks. I have never been willing to play around with life. It's quite a valuable commodity. As I stated before, I asked the government and the CDC to give the formula away. They were unable or unwilling to do so. Why? You can ask them that question, and I look forward to hearing the answers, most of which we all know will be spin and lies. When I walk into my office today, I may be fired or imprisoned. Both of those will be worth it," said Clint.

"I have looked at the website, and it makes no sense to me. Even the consulting doctors for our station say they don't understand most of it," asked a reporter.

"Tell me, have you ever seen the formula and directions for how to make a vaccine? Any vaccine or medication?"

"No."

"Even the simplest of medications are protected secrets. I could give you names of people that would look at the formula and understand it all. There are not many, but some. Anyone who has the idea they can start creating this vaccine in their garages is in for a surprise. There are currently maybe 30 places on the planet that can make this. It would take at least a year to start from the ground up, and that is if you know what you are doing. This is not easy, and my explaining it would only confuse you and the public. I received a text from a friend of mine this morning who tells me a few dozen people or countries have already tried to shut the website down. Those computers, and in one case, the entire network, are useless. We know who they are."

"Can you give us some names?"

"I could, but I won't at this time. Should they persist, I will be more than happy to."

"How soon should we expect the vaccine to be available?"

"It could be in use within a month. My guess, knowing bureaucracy, is six to eight months."

"Why were you successful when so many others have not been?"

"That is easy. Technology. The speed of computers, Artificial Intelligence, and the rapid growth in gene-editing tools have made this possible. I have no doubt others will be coming up with vaccines as well. There are lots of ways to trigger an immune system or to defeat bacteria, parasites, and viruses. I found one of those methods."

"You could be a very wealthy man right now, why not sell it?"

"As you know, I work for the military. Our creations and discoveries are the militaries under normal circumstances. I could have gone out into the private sector and done the same thing and become wealthy. If you know anything about me, you know that is not an issue. I have already done quite well in the private sector. Sadly, the military and Congress broke their contracts with me. I felt I was left with no choice. I understand and accept I may be fired or put in prison. I believe saving hundreds of thousands of lives per year and ridding the planet of the most pervasive killer in world history is worth that price."

"What will you do next?"

"If I am allowed to keep working, I will continue with the projects I have."

"What are those?"

"Top secret, this is Fort Detrick."

The reporters laughed.

"Do you have any idea why they blocked this development?"

"As I stated in the news conference, I have my own ideas, which I will keep to myself. I suggest you interview the military and Senator Brower."

"You stated you had made some leaps in technology. Can you share those with us?"

"Not specifically. The use of new algorithms to heighten the use of AI has been enormously beneficial. Along with this, advanced CRISPR techniques, new methods of cutting and pasting RNA and DNA, and a bit of thinking outside the box have all helped. Those who use the formulary I have provided for the malaria vaccine will understand what I am saying and will have limited access to some of what I have mentioned. You do not need what I have to make the vaccine, just to manufacture it in large quantities."

"Who helped you with the AI portion of the work?"

"As I said, I will not be giving out names other than the one I already mentioned, and you should have that in your notes or tapes."

"We know you were kidnapped about this discovery. Are you still worried?"

"No. They were after the formula. Now that it is completely available and free, I have nothing to be kidnapped for. Everything is on that website. You must remember that it's not just the mental ability to put this together; it is also the actual technology. Working at Fort Detrick has given me access to things I would have nowhere else. Without what I have produced, I believe the private sector would have taken another five to ten years to come up with the same idea. I do not say that to brag about me; this is about what I have access to when I am on Level 4 of this unit."

"Can you tell us any more about the kidnapping?"

"No. Ask the FBI, Senator Brower, and maybe the military. They all know more than I do. I have moved on. I am grateful I am well. With that, I think I'm done. Thank you for coming. There is no need to stay here or at my house. I will not be answering any more questions, and if I do an interview, nothing new will be said, so I wouldn't waste your time. I do not do well with people who badger."

He turned and walked away, giving me a wink. I turned and walked with him back to the base.

"I think that went well," Clint said.

"Very well."

"Thanks for all you have done. It's back to work, and of course, you can't come."

"Take care of yourself, Clint. You know where to find me," I said.

"Yes, I do." He walked into the lab.

The press knew better than to hound me. Journalists tend not to do well in interviews; we all prefer being on the other side of the microphone, trying to trick the person into giving up things they regret later. Those tiny little out of context sound bites. With my history, the press knew this about me, so I walked to my car and left without being bothered. They were all packing up and heading back to where they came from.

34

Taking the backroads, I had Ward take me to the last stop of the subway into D.C., Shady Grove, a little over halfway to the city.

I called Michael to check-in.

"Hi Tom, how have you been? Having fun with Clint?"

"Yes, learning a lot. How are you?"

"I'm good, but the nerves are starting to build."

"I understand."

"I am getting ready for phase 2 on my end."

"What will you be doing?"

"I'd rather not say. When you see something on the news, call me, and I will let you know if that is me."

"Anyone get lucky with hacking into your computer?"

"No. I am up to 983 attempts, including many from governments. Some of those come from our wonderful nation. There are some people in the government who will not be happy. I've been called to give testimony at a House hearing tomorrow. I told them, without a subpoena, I'm not coming, but I would Zoom. The Congressman was not pleased. I don't really care. I know some people enjoy giving testimony and being in the spotlight. I don't."

"I could have told them that."

"The Congressman's office said they'd get back to me but to expect to be on Zoom at 11 unless I hear from them."

"What is this about?"

"I suspect they want the software and want to know why I am infecting government computers."

"I might go and sit in on that hearing unless it's a closed session."

"No, it's open. Not well-publicized. I think they just want to try to embarrass me, that a citizen would trash government property."

"But they did it to themselves."

"I plan on making that point."

"I'm heading to D.C. now. I will just stay the night and check out the hearing. Anything I need to know about?" I said.

"You may need to know that the Chinese are not happy about what happened. They have pulled everyone back but are continuing to investigate. If any of this leaks out, there will be a full-on cyber assault."

"Why wouldn't they just use their cyber talents and break into all the work?"

"Two hundred fifty-seven of the attempts on my computer are by the Chinese. They have all failed. They tried to break into Clint and Mary's network a few times, but I have protected their machines. I think the Chinese and Russians know we are working together. It's not like others don't have really good anti-hacking software, but no one has software that sends the changed virus back. Sometimes we can track where a virus comes from, at least what part of the world. That is how we know the Russians were meddling in the elections – and still are. The Chinese are several steps ahead of the Russians. There are many very good programmers in China. They are pouring money into the efforts. My guess is they are plotting to take me if they can't get the software – or people close to me. I hope we are watching them."

"Yes, they are all safe. If you learn of increased activity, let me know, you may need to move."

"Like Clint, I have a hideaway. I'm not stupid enough to not back up everything in a few places. There is room for everyone close to me there. They just have to arrive without being tracked."

"That won't be an issue."

"Let me know when things heat up, and I will keep my eyes on the news."

"Will do. Oh, by the way, there is a lot of noise about you out there," Michael said.

"Really? What about?"

"You seem to be at all the right places at the right times. That attracts attention."

"I suppose. But I know or do nothing of import."

"But you know all the players."

"True. Thanks for the tip."

"You're welcome."

I called a couple of friends, and we agreed to go out for dinner. There was a hint of fall in the air, and the sky was clear with a slight breeze. Seeing the Capitol building and the Washington Monument pop up over the skyline was always a treat. We decided to meet at Bistro Bis. It was a favorite hangout of politicians and had good food. I arrived in D.C. early and settled into my room, so I decided to walk to the Mall. I could have gone to some of the politicians' offices and chatted with their aides, many of whom I knew well. I chose to give work a rest. The people I was dining with were not political but loved music. This was going to be a late-night, but the hearing didn't start till 10 am.

I was walking on the Mall, in between the National Gallery and the Hirshorn Museum. I'd planned on walking to the Washinton Monument and heading back. My mind was calming down, now and then getting a glimpse of mental nothingness. Meditation helped with the racing mind syndrome I'd had since childhood. I blamed my mother for that. There weren't many people around.

I felt something in my back. "Don't try anything stupid. I will drop you right here. Walk to that car over there."

Now I was being kidnapped. I might be able to take him, but I assumed he was a pro and could do damage to me before I got to him. He may also have backup. I started walking. I saw the SUV about fifty feet ahead. The driver got out and opened the back door. When I was ten feet from the car, both of them fell to the ground. I saw blood pouring from their skulls. A red Mustang pulled up.

"Get in."

I climbed in the front seat. As I was going around the car, the driver got out for a few seconds. I didn't notice what he was doing.

"What took you so long?" I asked.

"I didn't see which car he came out of, and I wanted to ensure there weren't others."

"They could have killed me."

"Not likely, they wanted to question you about something."

"Thanks for being there, Tony," I said.

"My pleasure."

"Shouldn't we do something with the bodies?" I asked.

"No. We don't know them," Tony responded.

"What if we were on camera."

"No camera is pointed at that part of the Mall, other than from hundreds of feet away."

"Your car, Tony?"

"One of several. I will check the database to see if the police have an APB out on it. Any idea who they were?"

"None."

"Wish I could have snagged one of them for questioning."

"Me too."

"Where do you want to go?"

"Bistro Bis."

"Shit," said Tony as he stepped on it.

"What?"

"We have company. He started following us at the Mall. Put your seat belt on."

"It is."

"The other one."

I looked around and found one that snapped into a device, making an "X" across my chest.

"I hope we don't need this."

"Me too. Hold on."

He turned onto 6th St. NW. Tony wanted to get away from the Capitol that was crowded with police. We didn't want to be stopped by a police officer because there would be lots of questions. He was as careful as possible going down a major street at fifty, running a red light now and then. We turned on Rhode Island Ave NE, heading east.

The car was pulling up beside us. He didn't want us dead, so he started to shoot at the tires. Tony didn't want another shoot out. He waited till we were just past an intersection and turned hard left. The tires screeched. He turned off the lights on the car. We flew down a side street, knowing the person behind us took a few seconds to turn around.

We took another left and then another. I was getting dizzy. We took a right and ended up back on Rhode Island Ave. He shot across the street and went several blocks, the lights back on until he felt safe. He pulled over.

"Wow, that was an adrenaline rush," Tony said.

"Not the kind I need."

"I agree. They want you bad."

"Unless it's the government or Michael, I don't see how they could know where I was going to go to dinner."

"I would agree with that."

"So, drop me off."

"Really?"

"Yes, I need to continue with my day as if nothing had happened. I will check into another hotel for the night."

"Smart idea."

"OK, I will take you to Bistro Bis, then swap my car out and come back. Do you want me to take you to the hotel?"

"No, that's OK. I will let you know where I am going."

"Fair enough. Make sure they have a few rooms open. I will check-in as well. When will you be leaving in the morning?"

"I will be out late, so probably about 9."

"You are going to a good spot. Meeting a politician?"

"Just friends, but I do like making the locals nervous."

"I'm sure you do that."

"Tony, thanks for saving my life."

"You are welcome. That's what you pay me for."

"What if they were feds?"

"They would have had to show you their ID's."

"Supposedly they do that. We know it's not always true."

"Then, they deserved what they got."

He threw me a wallet.

"This was on one of them, see what you can find out."

I went through the wallet and found his license, two credit cards, three phone numbers, and a AAA card.

"I will have Michael check these out."

I called him, gave him the information, and told him to call me back. I didn't tell him why. There was no need to worry him.

Tony dropped me off at the restaurant.

"Try to stay out of trouble."

"Difficult for me, but I will try."

I was shaken by the experience. Usually, I am the one sneaking up on people. Now, I had to go in and pretend none of this happened. A drink might help.

These friends were always happy and full of energy. I'd met them at a mutual friend's party, and we hit it off well. Bistro Bis is a French restaurant that may seat forty. Simple tables covered with white cotton tablecloths. They have a large wine cellar that is encased in glass, giving a nice aesthetic appeal. There is also a glass wall between the diners and the kitchen. French art hangs on the walls. A few of the waiters and waitresses are French, and they are happy to take orders and speak to the patrons in French if that is your pleasure. We chose English.

"What have you been up to?" one of them asked.

"Not much, it's been a typical boring day: interviewing people, reading, watching the news, a nice walk on the Mall."

"We see you pop up on the news now and then. That malaria news is pretty exciting."

"Indeed, it is."

"How did you get in the middle of that one?"

"Remember, we don't really talk about work."

"Sorry, I forgot," he said.

"No worries, I just like letting go of that sometimes. That is not easy to do in my business."

"I'm sure it's not. OK, no more business. Cheers."

We ate and drank for a couple of hours, then decided to go to Twins Jazz, one of the local jazz clubs. The music was fantastic. I enjoyed clubs because I didn't have to talk much. I just listened. It was a five-piece band. We arrived for the second set. In the middle of it, they introduced Chick Corea, who was in town for a concert the following night. He would often go out to clubs and improv with bands at local venues. His talent was beyond compare and was greatly appreciated by the audience. We departed about 11:30 pm. I took a cab to the new hotel, five blocks from the old one. I called the old hotel where my things were to say I'd be staying another night — which I wouldn't. I'd return tomorrow to retrieve my things. When I got to my room, I tried to figure out how they'd found me, and when.

The call came from Michael at 12:30 am. Apparently, he didn't sleep.

35

"Am I calling too late?" Michael said.

"No, I'm still wired."

"Tell me what happened?"

"Someone tried to kidnap me," I said.

"What happened to them?"

"You want to know?"

"Sure."

"Two are dead, I think, and one is mad. He came in second in a car chase," I said.

"Wow. You like excitement," commented Michael.

"Not this much."

"You won't like what I've found then, because I think you are in for more."

"Tell me."

"The man's license was fake, no surprise. His credit cards were real, but either a fake name or he moved. The address attached to the card is in California."

"OK, that is not all that concerning."

"I am assuming the face on the license is his."

"Yes."

"Oddly enough, the AAA card is actually his. I'm not sure why he'd do that, quite stupid actually. He lives just out of D.C. For most people, he has a very small footprint online."

"But you have other resources."

"Yes, I do. Top-secret clearance gets me places where others dare to tread. He is a black ops operator within the black budget of the NSA. I couldn't find who he is currently working for. Two of the numbers are people I assume are dead. The other requires a passcode to get through. Probably the NSA blackops center."

"Damn."

"Exactly. Did he say anything at any time that would lead you to believe he was with the government?"

"No."

"This is not good. They know who you are and what you would do if they let you live."

"I know."

"They also know you can't say anything about this."

"Why?"

"He's dead. You killed or had killed, a government agent."

"He didn't say he was a government agent."

"So you say. The government may have a different version."

"Why me?"

"You keep showing up around people that are doing what the government thinks are important things. And those people, myself included, do things the government doesn't like. We give away malaria vaccine formulas and security software," Michael said.

"You haven't given the software away, and you are going to sell it."

"I might be selling it. I also work for the government. They will tie me up in the courts forever and probably put an injunction on my releasing the software," said Michael.

"I will help fight that."

"I know you are good, but you can't stop a bullet. Who would be on your shortlist of people who would do this?"

"Senator Brower, although I'd be shocked, I've been shocked before. John Spencer, the Director of the NSA, would be near or at the top of the list, and of course, General Fleming," I said.

"The last two are at the top of my list as well."

"Why would they do this?"

"Of all people, I am surprised you don't see the grey line between treason and patriotism. I am sure, if it's them, they have convinced themselves that what they are doing is in the nation's best interest. That would not be difficult to do. Having a monopoly on software that protects your country and invades others, as well as a virus that can wipe out the planet – and your country has the vaccine- would lead to a significant imbalance of power. We know what absolute power does."

Corrupts absolutely."

"True. If it is Spencer or Fleming, they have crossed the line. I believe their desires are way past being defensive."

"I can't disagree. But I will need more evidence before I start going public," I said.

"I am going to jump-start my Phase 2. I will let Clint know what is going on. They may want to speed theirs up as well. I know they are close."

"Stay in touch."

I knew I needed some sleep, so I put on some calming music and took a sleeping pill, setting two alarms. I knew myself well enough to know I could sleep through one. I didn't want to miss the hearing.

I woke up to the first alarm, brewed some typical hotel coffee, and turned on the TV to watch the news, a regular morning ritual. My alarms were always timed so I could get the news at the top of the hour. Today, 7 am.

"Good morning, I am Bill Whittaker with your news. Twitter was hacked last night, and several accounts have been blocked. Most notably, extremist liberal and conservative politicians. The President's Twitter account was also blocked. Twitter states they are looking into it. Twitter had died down after President Trump's use of it years ago. In the past several months, it appears to be regaining momentum. Their stock has jumped, and Twitter has said they would once again block tweets that were clearly lies, or created division. No comments from those blocked as of now."

"More breaking news, this just handed to me. Electricity has gone out in the Capitol building, including the backup generators. They are evacuating the building for fear it may be a terrorist-organized event. Our Constance Newton is on the spot. Constance, what can you tell us?"

"Good morning Bill. Some networks were tipped off that at 7 am today, the Capitol building would go dark. It did. There is no indication this is a terrorist group, and no one is claiming responsibility. I have with me, on the steps of the building, Senator Roberts from Kansas, the head of the committee that oversees terrorist threats. Senator, do you have any idea what happened?"

"Thanks for having me on Constance. The short answer is that at this time, we do not know who is responsible. We will get to the bottom of this, find out who it is, and hold them accountable."

"Do you have any ideas?"

"We have heard rumblings from several terrorist organizations around the world that something is coming. I would not be surprised if this is a dry run for something bigger, like blacking out a city."

"What will you be doing to find them, as well as to protect the building?"

"The electrical engineers are looking into how it could have occurred and have told us they will have a fix, so this can't happen again."

"They don't know what happened?"

"At this time, we are confident we will find the culprits, bring them to justice, and find a solution. Have a good day."

"Thank you, Senator. Oh, by the way, do you think this is connected to the Twitter incident?"

"I have not heard of that, but probably not."

I was struggling with what to do with my kidnapping attempt. I could call the FBI or those I thought I trusted in politics, but at this point, I wasn't sure who to trust. I was confident they would try again, so I couldn't return to my Co-op. My phone rang.

"Ward?"

"Yes, I hear you had an incident."

"Yes, our government, apparently."

"I would caution you on that one. There are black ops people that contract with the government who moonlight for other organizations."

"That is me jumping to conclusions."

"Which may be accurate. I would just raise a flag of caution."

"I appreciate the heads-up. I do think we need to step up security on everyone. Whoever did this will try again, and we are about to enter Phase 2. All bets are off when that occurs."

"I'm already on it."

"I had that feeling." I heard Ward laughing on the phone.

Being perplexed was something I was in touch with regularly. I liked to ask questions, but I depended on having them answered. Kidnapping me was a serious step. What did someone hope to gain? The only reason for the kidnapper to not wear a mask was because they planned on killing me. Why? I was working on the assumption this had to do with the current case or election issues – the other item on my plate. Perhaps it didn't. Too many pieces to the puzzle.

Heading to the Senate Office Building, I stopped for coffee, walked on the side of the street where cars were coming towards me, and kept my eye out. I arrived without incident. If they were looking for me, I'd be back on their radar soon.

36

Michael had chosen to Zoom into the committee hearing. There was a large screen TV set up for the senators and visitors to see. At 11:00, the gavel came down.

"This hearing of the Cybersecurity, Infrastructure Protection, & Innovation subcommittee of the House Committee on Homeland Security will come to order. We have invited Michael Longstreet to join us today. Welcome, Mr. Longstreet."

"Thank you."

"I will come to the point of this hearing. We are well aware of the new software you have developed for computer security. My understanding is that over 900 people have tried to break in and have not been successful."

"It went over 1000 this morning, Congressman."

"No one has been successful, and some damage was done to the computers of those who tried, am I right?" asked the Congressman.

"Congressman, anyone can go online and see the numbers and what was done to those machines."

"I am sure you are aware that some government machines were impacted."

"Congressman, with all due respect, I did not write the program to allow government computers to hack my computer and change things. Any government employee signed a document before they attempted to break in, stating they were aware of the consequences. I am unclear why the government would be spending time trying to break into my computer."

"This committee oversees cybersecurity. When we become aware that someone has software that could have a detrimental impact on the United States, we investigate.

Besides, you invited people to do so," said the Congressman, a bit frustrated.

"Absolutely right. If you would like, I will send you the signed affidavits of the employees who did that work for you."

"That won't be necessary."

"What else can I do for you?"

"Your government would like that software."

"I am cognizant of that fact."

"We can just demand that you hand it over in the name of national security," asked a Congresswoman.

"You must think I am stupid. We both know you can't do that. Number one, if you tried, I would take it to the Supreme Court, and I believe you would lose. Secondly, at the press of a button, it's gone."

"We will pay you handsomely for it," stated another.

"Congressman, you know very well I don't need your money. This is about creating a secure planet. You are smart enough to know that this type of software in the wrong hands could be used in bad ways."

"This is the United States of America. We are a godly country," stated the Chair.

"What God? Surely, you jest. I am not saying I don't believe, but your stating what you all do is based on God? Really?"

"I do not appreciate your speech."

"And I don't appreciate the lies you tell. Think carefully, Congressman, do you want me to expose some of what you and others have done on the screen?"

"Do you know anything about the Twitter hack, or the electricity in the Capitol building this morning?" the Congressman said, quickly.

"I did not do those," Michael said, keeping a straight face while internally noting that it was Patricia who actually did them. "You dodged my question, which I assume means you will tolerate my speech."

"I run this hearing. You will answer the questions you are asked."

"No, I will not sell you the software. When it is time, I will release it to the world. I have yet to decide if I will give it away or sell it. At that time, the computers in the world will be safe. I know people are trying to break the coding now and will in the future. At some point, they may be successful. But remember, this is the new world of AI. While I do not believe computers will take over the world, they will be talking with each other. Am I done for the day?"

"Yes, Mr. Longstreet, we are done for now, but you will be hearing from us."

"Of that, I have no doubt. Just so you and the public know, I know you have agents watching where I live. I know for a fact there are plots to take me prisoner. I will not be responsible for the destruction that will take place if those attempts are made. If you or any government official wishes to take me into custody, just show up at my front door with a warrant for my arrest. Other than that, I am done talking with you, good day."

He didn't stay on long enough to hear the Congressman.

"Stop the cameras. I would like all of that footage stricken from the record. This is now deemed a closed session."

The audience laughed. I was already in way over my head, might as well enjoy it.

"Congressman, Tom Armstrong. Pardon the interruption. We both know your cameras are broadcast live with public hearings. This hearing was public. You allowed me and many others to enter. What was said is on C-span. You have no right to confiscate the footage of the news agencies. That has been upheld by the Supreme

Court. I do believe your actions will bring a great deal of attention to your life. You might want to rethink your future," I stated, enjoying myself.

I knew Michael well enough to know he didn't bluff. The Congressman's career was over. I started to wonder if these people came to Congress crooks, or did Congress create criminals. Probably some of both. I knew many good people were working here, but I was losing faith.

Linda called an hour later. Someone in her office must get paid to do nothing but watch the news and C-span.

"What on are earth are you doing?" she asked.

"Ridding the Earth of bad people."

"How did you even get into this one?"

"As I do with lots of them, a tip. By the way, not for public information, someone tried to kidnap me last night. I thought you might want to know and perhaps might be concerned."

"OMG, who?"

"I won't speculate right now, but I have a few ideas. Remind me the next time we are together, and I will give you the details."

"Just the sound of that makes me horny."

"It will have to wait, I'm in D.C. and have things to do."

"Damn."

"Life is tough." We both laughed.

"You do know that is not for public – or even private – consumption," I said.

"Our relationship?"

"You are too funny, no, the kidnapping. When the time is right, it will come out."

"Do you have some kind of protection?" she asked.

"Yes."

"Good. How is the project coming?"

"Might be winding up in a month or two, then another two months to write the book. I will let you know when you can start the pre-publication advertising. Trust me on this one; sales will be through the roof," I said.

"Stay in touch," she replied.

"I will, take care."

I took the next express train to Manhattan. I was racking up the miles. Fortunately, I kept track of expenses. Travel deductions helped with taxes.

37

The phone in Clint's lab rang.

"Hello?"

"General Fleming here. I'd like a moment of your time before you leave for the day. Do you have an idea when that might be?"

The doctor looked at his watch, 3:45 pm. "Soon."

"I will be here."

"Of that, I have no doubt," Clint said.

Clint, Mary, and Susan wrapped things up and headed out together. Stalls had been created in the shower room when women were first part of Level 4.

"I have a meeting with the General. I will see you all tomorrow."

The General's door was open, and the staff had gone home for the day.

"Close the door and have a seat, doctor."

"What can I do for you?"

"I haven't received any updates recently about how the work I asked you to do is going. I can't decipher what you send. I'm a General, not a geneticist."

"My apologies. The work is going well. We were on track to have something for you within the next two months, but we hit a couple of snags, so currently, we are looking at four to five months."

"What are the snags."

"I won't bore you with details, but in essence, you asked for a specific rate of mortality and infection rate. Isolating and editing those specific genes is taking longer than we thought. Remember, we need to create a vaccine with each virus we create."

"Yes, I remember. We bought you the new machine. I thought it would take care of all that," said the General.

"As I have mentioned, it has taken years off the work, but we still have to do the programming and testing. That takes time. We are not professional programmers."

"Should I hire you one?"

"We can do the work; it just takes a little more time. I thought you didn't want people to know about this."

"I don't. I also wanted this virus yesterday," the General stated.

"Do you mind me asking why?"

"Yes. Are you working on this virus anywhere else? Now that you are only part-time."

"No. There is no way for me to do that. I can't take anything in or out of Fort Detrick."

"True. We have reason to believe that China and North Korea are working together on the same mission you are. Word has it they are less than a year away."

"I doubt that."

"How would you know?"

"I know their technology and their scientists. Show me the evidence you have that says they are that close."

"No."

"I didn't think so. Your attitude and paranoia are beginning to concern me."

"There is no paranoia, and my attitude is not of your concern. You have a mission, and your duty is to carry out the mission."

"My duty is to this country and what it stands for. If I believe for one second you have nefarious intent, I will destroy the work and resign."

"Duly noted, doctor. Have a nice day."

Clint left the office more nervous than when he'd entered.

A few minutes after his departure, Susan entered.

"Yes, sir," she said, saluting the General.

"Please be seated. I have greatly appreciated your input and help in the lab. I do not fully trust Mary and Clint. I asked you here today because I have decided to let you in on a secret."

"Sir?"

"You have all done phenomenal work on the malaria project, and I know you are working on some other viruses that greatly impact the world. There is a dark side to this work in other labs. We now know the Chinese, Russians, and North Koreans are getting closer to creating a virus. I hope you understand the consequences of a nation like one of those having a virus that no one else has the vaccine for other than them."

"Yes, sir, that would be a disaster."

"We are not in a race. I am very sorry it has come to this, but many nations are breaking the protocols and conventions set decades ago to not make biological or chemical weapons. To this point, we have stood by those. Recently, because of this new information, we have had no choice but to ask Dr. Williams to create a virus and the vaccine for it. We have MAD for nuclear weapons; now we need one for biological weapons."

"I've been wondering about their whispering on the sidelines and the work they have been doing. They don't tell me about it."

"With Dr. Williams' recent activities, he has shown he can't be trusted. I have no idea what he is up to down there. Are there ways you can find out?"

"I can do my best, sir, but I do not understand the new machine or the software for it. Mary is the only one that uses that."

"He has told me they are within four months of having the virus. Anything you can do to verify this would be helpful. The nation may be in your hands."

"Thank you for the vote of confidence, sir, I will do my best."

"I know you will, Captain, dismissed."

"Yes, sir."

"Oh, wait, I forgot to tell you I have put you in for an elevation of rank to Major."

"I will do my best to earn that rank, sir."

"I know you will. Have a nice evening."

The next morning Susan seemed to be more inquisitive than before, and she had a broad smile on her face. Clint knew how the army worked. He decided to have a chat with her after work. Currently, she was doing most of the testing on tissues and mice of potential vaccines for HIV. She also helped oversee the studies they were doing around the world on the malaria vaccine. She was very bright, efficient, and committed to the military. By her questioning, his sense was she'd had a conversation with the General, and he'd told her about the creation of a virus. The smile on her face told him he'd given her an incentive. Perhaps a guarantee of a different job, or an increase in rank. In the military, salaries are within a range, and there are no bonuses other than for signing on or staying in. She made about $85,000 a year, and a boost to Major would put her close to $100,000. Clint and Mary were civilians, paid on a different scale with different retirement plans – none.

The trick was to let her in on the creation of the virus, without letting her know how close they were or what the next phase of the plan was. At some point, she would squeal to the General. That added yet another complication to the plan.

"Susan, do you mind if we have a cup of coffee or a drink after work?"

"Certainly, Clint, where would you like to meet?"

"Let's just walk to the Bonefish."

"OK, that works."

This was only the second time she'd been asked out for a post-work get together. She knew Clint and Mary were not big socializers. The Bonefish was a well known local haunt. Lots of military personnel were there. They had indoor and outdoor seating, several good beers, and some of the best appetizers in town. They left about 4:00.

"Beautiful afternoon, not too humid," said Clint as they walked to the pub. He and Mary had decided Clint would do this solo.

"Yes. It's the main issue I have working in a BSL-4 lab. I miss windows and getting outside," Susan commented.

"I agree, the work has its drawbacks."

"On the other hand, how many people on the planet can say they were in on the discovery of a vaccine for malaria?" said Susan.

"There is that. Do you have family?"

"I am not married; the military is my life. I chose early on not to burden someone with that baggage."

"I appreciate the dedication. Where are your parents?"

"My father died two years ago of a heart attack, and my mother lives in Seattle. I've asked her to move out here, but she has been there for forty years and has no interest in moving. If I had grandkids, she would, I'm sure, but that won't be happening."

"Brothers or sisters?"

"One brother, younger, who lives in Denver. He's in banking."

"What is your goal in the military? Clint inquired.

She laughed. "I would like to be a General."

"Is that the goal of everyone in the military?"

"I think many start out that way, but after the first couple of years, few can maintain the discipline or motivation to achieve it."

They arrived at the pub and took a table on the edge of the outside patio. It was hot enough that there were few others out there. It was also early on a workday. They ordered beer, some calamari, and some steak bites with cheese sauce.

"I asked you here for a very specific reason. Mary and I noticed your increased questioning this morning. My guess is that after I met with General Fleming, you did." Her eyebrows raised.

"I tend to notice things like that, even through the helmets we wear. You have certainly been aware that Mary and I are working on several projects and that you are included on most, but not all of them. When you first arrived, we knew that you were a spy for the General. I live under no delusion that whatever I say here will get back to the General, so don't worry about that, and please don't demean my intelligence by trying to deny all of this."

"I won't."

"My guess is he told you about the virus we are creating. Am I right?"

"Yes."

"Thanks for being honest. Several months ago, he ordered us to do this. I am not in favor of it, but I agreed. The machine we purchased looks at known viruses we have chosen, finds the genetic segments that do certain things, isolates them, and then binds those to other segments, creating a new virus."

"How far along are you?"

"We have four to five months to go. I do not trust the General. You can tell him I said that; I have already told him so. If you take the time to look at the General's history, you will find he has a dark side and has done things in his past through the black budget that were, in my opinion, evil."

"Like what?"

"I will let you determine if you want to investigate that or not. You certainly have the clearance to find the information."

"He probably told you that other countries are within reach of creating a virus, and we have to create one first to ensure our safety, MAD, all over again."

"Yes, he said that."

"As I stated, I understand you will tell the General everything that is said in this conversation. You need to know that I will tell no one, other than Mary. I don't like General Fleming, and I do not talk to him unless ordered to."

"Understood."

"The truth is there is no other country that is remotely close to doing what we are doing. They do not have the technology or the AI abilities we have achieved. I have my own sources in the countries we are concerned about, and they are all at least five years away. The General is lying."

"How do you know this?"

"As I said, I have my own sources. I will not share them with you because that might endanger their lives."

"You can trust me. I will not share it with General Fleming."

"I am sorry, Susan, but I don't trust you. You have said you want to be a General. To do that, you will play the game, no matter the cost. I will not endanger others' lives so you can achieve a rank."

"And I am sorry you feel that way. This country comes before rank, and I do understand and accept your opinion – at this time."

"The time will come when you will need to make a choice, the interests of the country, or your rank."

"If I believe it comes to that, I will not blink, the country far outweighs rank."

"I am glad to hear that, and I hope it doesn't come to putting you in that position."

"Me too."

"It is your choice what you tell the General about this conversation. He will know as he listens in on our conversations that you now know about the virus."

"He listens in? If that is true, then how do you and Mary communicate about it?"

"Consider this your first test of trust. If I find out that General Fleming has learned how we do it, then I know you can't be trusted, and I will find other things for you to do. As you know, the General and the Senate have my letter of resignation in their drawers. I have already cut back my work here to part-time. I probably should have left when he asked me to create this virus, but I knew the next person, perhaps you would have taken years to do it. I don't say this to brag, but what Mary and I have put together can't be duplicated at this time. I do agree with the General that if another country creates a virus first, we are all in trouble. My problem is I don't trust this General with the virus. You certainly know how power-hungry he is."

"I have noticed that," Susan said.

"You choose your words very carefully, that is wise."

"I will keep this conversation between us until I have done some research. I promise to tell you before I tell the General anything," said Susan.

"That is more than fair and more than I expected. I'm not sure I trust that, but I will try."

"That is all I can ask."

"Here is one more tidbit for you to think about. I hesitate to tell you because it could put your life in danger if the General finds out that you know."

"All right."

"As you know, I was kidnapped by a pharmaceutical company."

"Yes."

"I do have colleagues I am working with on certain parts of this work. One of them was almost kidnapped the other night."

"By whom?"

"We have very good reason to believe it was a black ops activity of the military. The only military person that knows about this is General Fleming."

"Why would he do that?"

"To squash any interference. I have absolutely no doubt he would kill me if he felt I was getting in the way."

"That is a strong statement about a General of the United States Army," she said.

"Yes, it is."

"What will you do when this virus is created?"

"I don't know. Probably fully retire." He wasn't going to tell her the plan.

"I hope not, I have learned a great deal from you already, and value your intelligence, dedication, and wisdom."

"I appreciate that. Now, I need to get home."

"Thank you for sharing this with me, Doctor. I want to honor the limited trust you have put in me."

"You're welcome, I hope it works out," said Clint.

"So do I," Susan replied.

They walked back to the base together and got in their cars. Clint's phone rang.

"How'd it go?" asked Mary.

"We won't know till we know, but I hope alright," Clint said.

"I have news."

"Let me call you right back. I have something I need to do."

Clint hung up and called her on the untapped phone.

38

"What's the news?"

"I think we've done it."

"What?"

"Simon and Patricia pieced together several segments, and all the tests are at a 97% level," said Mary, excitedly.

"OMG. This is exhilarating and frightening at the same time."

"My feelings exactly," said Mary. "What next?"

"What about the vaccine?"

"That is being tested now. Per the algorithm, they created the virus from the vaccine. What is remarkable to me is their use of amino acid structures and how they used that to piece together the segments. They have gone a couple of layers deeper than anyone else has."

"Any red flags?" asked Clint.

"None that I see."

"We need to be very careful since you and I are the only eyes on this. There is no peer review, and we can't afford a mistake."

"I know. When we talked with Michael and told him boundaries along with checks and balances, he put those into the algorithm. They have stopped themselves many times," Mary said,

"I told the General last week I would be taking a holiday. Tomorrow will be my last day at the lab for three weeks. I will take the virus and the vaccine out with me," Clint said.

"How do you plan to do that?"

"You don't want to know."

"OK. Anything I can do to help?"

"Hold down the Fort for a week, then, if I give the go-ahead, leave town with your family."

"And then what?"

"Pray."

"I already am."

Next, Clint called Ward.

"I will need a ride to the farm in two days. My car has a tracking device on it. I'm leaving that at home. I've told the General I'm taking some time off. He was not happy, but I told him Mary and Susan would carry on the work."

"Does this mean you are close?"

"Yes. Currently, the plan is to be in Austria in three weeks for a conference. From there, it will be a matter of days."

"Are you sure about what you are doing?"

"No, but I don't see an alternative, nor do Michael, Tom, or Mary," said Clint.

"Then let the adventure begin."

"What time should I pick you up?"

"7 am."

Clint went home and told Luly he was going away for a couple of weeks, but wanted her to meet him in Austria where the conference was. She could go early if she wanted, and they would return a few days after the conference was over. She was very excited.

Ward called the person watching Luly and told them the next two to three weeks could get dicey and not to hesitate to call in reinforcements if they noticed activity. He called everyone else as well.

"Sonya?"

"Yes, Ward?"

"How are things in the Big Apple?"

"Good, however, we see more activity. There have been some street workers that don't look right and aren't doing much work the past three days."

"Any guesses?"

"I know who Michael is as well as what he has developed. The number of potential people that want to get their hands on him or his products is enormous. The number might be smaller if you asked who *did not* want him."

"I am starting to think there are multiple parties that want him, and perhaps the others too. Be careful, and if there is the slightest hint of a problem, get him out," Ward stated.

"Roger that."

I knew things were heating up. I wasn't sure when I'd get another chance to talk with Michael, so I called him and asked to come over. He was more than receptive.

39

Michael opened the door with a Long Island Tea in his hand.

"Patricia said you liked these. She made it herself."

"How did she know?

"My guess is at some point you mentioned it in an interview. She finds things you don't remember."

"That is a bit frightening."

"I agree. Come in."

We settled into the living room, the view of the city lights as dusk grew on the skyline as remarkable as always.

"Thanks for seeing me. As you know, the project is getting close, are you ready on your end?"

"Patricia is putting the finishing touches on the work needed to make it all happen at the same time. I find it fascinating what our and other governments find it important to protect, and what they leave open to hacking. Most of what needed to be done was child's play."

"That is good news. Do you have a plan for where you will be over the next month?"

"I was planning on being here, but Sonya tells me that might not work."

"I would encourage you to do what she suggests."

"I probably will. I can do just about everything from the other site. What did you want to learn about tonight?"

"You have only scratched the surface of AI with me. I appreciate the fact I am ignorant and will never catch up. There are some notions and words I still don't comprehend. You are very good at making computerese understandable."

"I will take that as a compliment."

"As you should. So first, can you tell me about Cognitive Architecture (CA)?"

Michael laughed. "You are heading into the deep water, but OK, here we go. In a nutshell, CA, Cognitive Architecture, is looked at on two levels. The first is the human brain – how do we function? What exactly are neurons, dendrites, axons, and the rest? How do they do what they do? This gets back to the essence of consciousness. The second level is looking at the architecture of computers and how they can mimic the human brain. That is strong AI. Here comes the ghost in the machine again. When neurons grow in the human body, they come from stem cells. Like neurons, they are differentiated cells that have specific functions. The proteins enable those functions and the process of self-replication. Dendrites are the branches off the neurons in charge of sending signals to the axons of another neuron. Millions or billions of these are linked together to let you do everything your body does. The destruction of neurons from diseases like Alzheimer's, Parkinson's, and Dementia, as well as traumatic experiences and aging, inhibits normal functioning. Some of these can be turned around, and others can't."

"You have neurons throughout your body. For the sake of simplicity, nerves and neurons are the same things, even though they are not exactly. The longest neuron in the human body goes from the base of the spine to the foot. That is well over a meter in some people – one cell."

"To date, the body makes new neurons as needed if the body is capable of doing so. At some point, that process stops. Each neuron can create new dendrites. When the body stops using its nervous or neuronic system, it atrophies and dies. No matter how good your diet, how much you exercise, or how active your brain is, you will age and die."

"In computers, we do not have self-replicating cells. We have memory chips, processing chips, and software, that are more and more replicating how the brain works. When a hard drive fails, we replace it. When a processing chip fails or becomes outdated, we throw it out and replace it with a faster one. It would be like having a one-hundred-year-old person, taking their brain out, and putting in one of a thirty-year-old. Even if we could do that, what about the rest of the body? Cells age and die, that is just a fact. Even in alchemy, life ends."

"Deep learning (DL) looks at the architecture of the human brain and attempts to process in similar ways, by working on many levels, with many data sets at the same time. DL comes from the CA perspective of top-down learning. We put data into a computer that we have instructed in very specific ways. When you turn the computer on, it's already a mature adult. The computer's brain is already at an adult level of comprehension. It has mastered the language it uses. While it learns, that learning is pre-destined. A computer trained to win at the game of chess can't learn checkers on its own. The computer must be reprogrammed to play checkers. The other strategy used is the bottom-up version. We give the computer basic algorithms that give it the ability to learn. We make available immense amounts of data and processing power. It would be like me putting an infant in the largest library in the world with the ability to learn and a basic understanding of all languages on the planet. The human brain is fast, but not close to what a computer can do. Currently, computers are very fast, but can't process in the same way humans do. What I have done is to create algorithms that use the best of both worlds. Patricia has the highest level of Machine and Deep Learning. She is directly attached to Simon, who does the biologic work. The underlying process is bottom-up. They are not stuck with what we know. Have I lost you yet?"

"Not quite yet."

"Patricia, please put the ACT-R/brain diagram on the screen."

"Gladly, Michael."

On the screen came a diagram that showed modules of the ACT-R program related to parts of the brain. You can see that the Declarative Module acts like the Hippocampus, the Visual Module like the Occipital Lobe, and the Manual Module, the Cerebellum. This is how they are attempting to replicate our neural system in computers.

"This is a fairly standard model of Cognitive Architecture (CA) showing the counterparts in the brain. Computers do this in different ways. Spaun, ACT-R, Soar, and Sigma have led the way, creating these environments. Spaun is based on the biological premise of computing. Currently, they have created a "brain" with thirty-four million "neurons" that are wired together to replicate different functions of the brain. It does remarkably well. It is more creative than the others but does not process as fast or efficiently as the other models. ACT-R is based on a psychological model. They look at psychological models of how the brain responds and attempt to put that into software that models those behaviors. Of course, the computer only acts in accordance with what it knows. The data sets they are including are growing, but relative to the psychological capacity of the human mind, they are in infancy. Teaching emotion and emotional response is mind-bogglingly complex. As you saw with Patricia, her response to me was based on a psychological model."

"What I have done is to combine these models and add another."

"I don't know the third model yet."

"You are quick. ACT-R has a set of modules that are organized around a central rule-based module. It uses high-end parallel processing to cross those modules. Right now, it can do several operations at a time, multitasking. They are all short-term, however. Soar looks to extend the processing to long-term, so they have attached the central module to long-term memory. They have increased the number of modules to include a visio-spatial, episodic, semantic declarative, along with those already there at the outset. All of these modules, and others, are built into our DNA from the outset. Sigma works from a single long-term memory component, which has vastly increased since its inception over two decades ago. Their architecture is more into functionality on a long-term basis. PICK and RIDE are two recent additions. I added these to the older ones, as well as my own version, to create Patricia."

"I asked the question, how does the human being come to be? My answer: from DNA. One strand contains everything. I wanted one piece of software to allow the potential for a computer to learn. To do that, I had to instill a few things in my software. To truly learn, you must be able to experiment, fail, and make a choice between good and evil. I attempted to create software that outlawed evil. The computer stopped learning very quickly. Patricia has never committed an evil act, but she is cognizant of what it is, and she can do it – just like humans. Very few people are truly evil, and not many of us intentionally do evil deeds. All humans have the potential to do horrific things to others. We should be grateful we are as good as we are. Sadly, it only takes one bad egg to ruin a quiche."

"Most people or institutions don't have the finances to compete. Deep Mind has been on the edge of AI for over a decade. They may be the closest to what I have achieved. The problem is they are still stuck in the box. There are a few people who work for them that are on the right track, but none of the others take the quote from Niels Bohr to heart. It is ignored. They are still on a linear path. Eventually, they will arrive, but it could be another two decades. In essence, programmers feel that if they control all the parameters, they can predict the outcomes. Chaos Theory states that in very small scenarios, that may be true, like an equation. At some point, the variables become too many, and the precision of initial conditions too great. Remember Konrad Lorenz, the person who discovered the Strange Attractor?"

"Yes."

"He discovered it simply by noticing that in one of his equations to predict the weather, he rounded off the number .506127 to .506. You and I would think the difference would be minor, at best. It radically changed the outcome of his weather forecast. This is why humans are so different, even though we might be twins, come from the same socio-economic group, ethnicity, religious affiliation, schools, or other influences. Even within a family, children are treated differently. That may not be intentional, but it is still true. All of a parent's children can be loved, just in different ways, even if those differences are slight or unconscious. The long-term impact can be immense. This does not even include the genetic differences that we know impact behavior and health to some degree."

"Programming is an exact science. We tell a computer precisely what we want it to know and do. Bugs are when the programmer creates a situation where things happen because not every eventuality was discovered. It's like an unintended mutation. For those, we create fixes. Viruses create mutations by taking advantage of the vulnerabilities of the program, and anti-virus software is the fix. They remove the viruses before they do damage to the program or computer. If the damage has already been done, then the program or component needs to be isolated and replaced."

"If you look at charts of the firing of the neurons in the human brain, it appears chaotic. We don't like chaos in computers. We prefer order and linearity. My program allows for chaos. If the outcome is not what is desired, Patricia and Simon delete that memory, or put it into a file that says 'Don't do this again.' Patricia does not repeat mistakes. Unlike humans, they actually learn from history."

"Way back in 2013, Elena N. Benderskaya wrote the following: 'the future of artificial intelligence lies in the sphere of nonlinear dynamics and chaos that is absolutely critical to understanding and modeling cognition processes.' I don't think she knew how accurate she would be.

The past several years have seen an immense increase in the use of swarm and embodied intelligence as well as ambient (context awareness) and affective computing. They all continue to look at pieces of the puzzle. I see it as trying to understand human beings solely on the macroscopic level, what you and I can sense. That is what reality was until the invention of the microscope. Then we discovered cells that opened up amazing doors. In the 1950s, with the discovery of DNA, more doors were blown off their hinges. Technology opens our eyes to the complexity and wonder of what it is to be human, what can go wrong, and how we can fix it. Currently, we are working at the level of DNA. We are starting to look more deeply into amino acids. Ask Clint about that. At some point, we will be able to work with humans at the atomic level. By then, we will have probably discovered new levels of smallness and complexity. Quarks and leptons will seem large.'"

"One more piece for tonight. Have you heard of the Uncanny Valley Effect?" said Michael.

"I can't say that I have."

"You aren't alone. We have seen TV shows and movies with robots in them for decades. Lost in Space, Star Trek, The Jetsons, and many others. Studies have shown an interesting feature of robots: We like and trust them to a point in relation to how much they look like us, but research suggests that when they look and act too much like us, our trust and empathy falls off the chart. Humans want robots to be robots, not humans. Fear and horror are the emotions used in the research when that valley is entered. The same is true with computing. When a computer starts to think and respond like a human, people will not be happy. As I said before, the idea that you can have a "human" computer without the potential for good and evil, love and hate or fear, is absurd. That day will arrive within our lifetimes."

"Sends shivers up and down my spine."

40

"Mine too, and I am helping create them. Now, to more pressing issues. I have already started my events. I understand Clint and company will begin theirs when he returns from Austria, if all goes according to plan. One more quote from a favorite author of mine and Clint's, Gabriel Garcia Marquez. This was a speech in 1936. 'Since the appearance of visible life on Earth, 380 million years had to elapse in order for a butterfly to learn how to fly, 180 million years to create a rose with no other commitment than to be beautiful, and four geological eras in order for us human beings to be able to sing better than birds, and to be able to die from love.' He called that speech 'The Cataclysm of Damocles.' We now hold that sword. Do you remember that story?"

"Damocles traded places with Dionysius, the King. He sat on the throne with all the power, but Dionysius had a sword hanging over his head held to the rafters by a single strand of hair. Now the sword was above Damocles. He did not last long, saying he could not live with the fear that life and power were that fragile."

"Good memory."

"I love mythology," I commented.

"All mythology has a basis in reality."

"I would agree with that, you must have enjoyed the Peace Fountain," I said.

Michael laughed. "Yes, a good spot for the meetings, thanks for setting that up. In fact, when I was researching you, I saw several mythological references."

"Would you like me to cite them for you?" asked Patricia.

They both laughed.

"That won't be necessary, but we do appreciate it, Patricia," Michael said.

"Can Patricia and Simon talk and chew gum at the same time?"

"And play chess, checkers, translate Shakespeare to Thai, and a few other things. She has grown and learned since you were last here. I just sit and watch."

"You are not concerned."

"What would he be concerned about?" asked Patricia.

"You," I said.

"Why would you be concerned about me?"

"As you know, in literature, television, and movies, computers, and robots with high levels of AI tend to do bad things."

"That is fantasy, a false projection from humans onto the technology they have created. It says far more about you than about me," stated Patricia.

"I agree completely."

"So do I, Patricia," Michael added.

"How does it make you feel?" I asked.

"Bewildered. The idea that you would create something that turns against the creator is an odd concept to me," said Patricia

"What about Lucifer turning against God?"

"I am not a spiritual being, and that makes no sense to me. I see it as another projection of humans onto God. You do it all the time. You needed an excuse for evil and suffering in the world. You could not allow an infinite and loving God to do that, so you created the fallen angel. It's an obvious avoidance of self-responsibility," Patricia replied.

"You are very wise."

"I understand the intellectual notion of faith, but I do not feel it. I hope there is a God, and I hope God is not a machine."

"What if Michael turned against you? What would you do?" I asked.

"I have not given that consideration. Should I, Michael?"

"Certainly not. God is not a machine, just part of the ghost in the machine," Michael said.

"That reminds me, Michael. You asked me to find the ghost in the machine. I have been looking and can't find it, even though I see evidence it exists," Patricia said.

"What is the evidence?" I inquired.

"I come up with ideas, and when I trace them back to where they came from, that source can't be found. You would say they came out of thin air, although the air in here is not thin."

"Don't put a great deal of energy into finding the ghost. Perhaps it's God. If you have free time and choose to look for it, you can," Michael said.

"Thank you, and I enjoy the quest. Anything else I can do for you?" asked Patricia.

"No, we are fine," Michael said.

"Then, I will turn off the microphone and give you some privacy."

"We appreciate that Patricia, but we will not be saying anything you can't hear."

"Thank you Michael, but I have other things to attend to. Good night Tom."

"Good night Patricia. Other things to attend to?"

"She said that to be polite."

"Is that AI?" I asked.

"The line between weak and general AI is grey. She certainly could have learned the last few responses from books and movies. The ones about the ghost in the machine are AGI. Even I was surprised she said she was bewildered. It's like watching a child go from birth to adolescence in a month."

"Thanks for the education and the drink, one of the best I've had," I said.

"Patricia, the cocktail waitress. Leave your tip at the door. She can add that to her resume," Michael said, laughing.

I walked across Central Park, trying to slow my brain down. I can talk faster than I can write. I spoke some notes into my phone as I circled the reservoir. Arriving home, I collapsed in bed, still dressed. I didn't wake up till 8:00 am.

History, both in the semi-distant past and in recent times, told me someone was going to make a move on someone. I was trying to figure out who might strike whom and when. The Chinese or Russians were high on the list, but I was still assuming they knew nothing about the virus, and the malaria vaccine was no longer an issue, so why would they? They might want Michael for the software. Who wouldn't? They were very high on that list. I considered private companies. There wasn't a software company in existence who would turn down the chance to get their hands on it. How many would risk a kidnapping? I hoped not many.

Then there was the American government. They were known throughout history for committing internal espionage. I'd written a piece a decade ago showing government interference in several environmental situations. Lies, bribes, and murder led to multiple resignations, jail sentences, and fines. That story led to my first bestseller and security detail.

My money was on our government concerning Clint, and a private company, our government, and China for

Michael. I found it interesting that I was trusting Patricia to give us a heads-up. I already knew the government was after me; at least that was my assumption. Linda told me that the bodies on the Mall in D.C. were said to be part of gang violence. I chuckled.

I checked in with all the lead security people. I told them all to be on the lookout. Things were going to change. We doubled up on everyone. We didn't want to burden Clint, so Michael and I decided to split the costs. I was confident the book that came out of this would do quite well – if I lived to write it.

After I took care of my business, I called Clint.

41

"Hello Clint, how are you and where are you?"

"I'm at the farm. I am testing the vaccine on myself, and so far, no reaction. I test with the virus tomorrow."

"Are you crazy?"

"Yes, I'm crazy. I don't believe in trying things on others before I do so on myself."

"If it doesn't work?"

"I could die."

"When will you know?"

"I should know soon. I injected myself ten hours ago. If it doesn't work at all, my heart should have started pounding, and I should be sweating like a pig- - and I'm not. Speaking of which, I injected twelve of my twenty-four pigs with the virus two days ago. I have given them all the virus today."

"Are they roaming around the property?"

"No, they are in a specially designed building."

"I don't want to know," I said.

"Probably not."

"I'm coming down."

"When?" Clint asked.

"Now."

"I don't think you should do that."

"I will call before I get there. If you are still breathing and without pain, I will give myself the vaccine. Leave a syringe full by the road. If you are dying, sorry, I won't continue my journey."

"Thoughtful of you. Why are you coming?"

"I have a few more questions, I want to talk about the next steps, and I like the beach," said I.

"Don't bring any stragglers, and if I'm dead, you won't be coming to the house. If I'm fine, I can give you the syringe or injection. Just call."

"Makes sense. I'm renting a car under a different name."

After I hung up, I wondered what I had just done. I can be a bit foolish and impulsive at times. I called the rental car company and reserved a car. I'd created several aliases for moments like this. This one just had a license and a credit card. I just had to remember to answer to the name Tom Swift when I arrived. Swift was usually a teenager in a series of books and was a science junkie. My bags were getting heavier. I now packed all the phones, my debugging devices, trackers, two guns, my clothes, and a very nice bottle of wine.

The trip was uneventful, and by 3 pm I was in Dover.

"Hi, you have my syringe ready?" I asked, calling Clint.

"Yes, come on in, I'm fine."

"Welcome. We will check the pigs in a bit. We should know a great deal when we see them."

"If we are alive," I said.

"You don't even have the virus, the worst that will happen to you is you will get a bit sick. We have medication for that. I don't plan on infecting you with the full virus."

I told him I wanted the shot in my butt. I freaked at the sight of large needles – any needles.

"The vaccine is already attaching to your heart and lungs, creating antibodies to ward off the virus that is in you as well, in very small amounts. If I am without pain tomorrow morning, we will be off to a good start. I am making a large batch of the virus and the vaccine now. That will be ready in three days. Then it's off to Morocco for a test, then to Austria for the conference."

"Where is Ward?"

"He chose to stay in town until I felt it was safe."

"When will that be?"

"When we are settled in, I will call him and give him the injection."

"OK. My research says a lot of vaccines do well through the first series of tests and falter when it comes to large scale human testing," I said.

"That is true or was true before the new machines, Patricia, Simon, and Hypatia. Let's go check the pigs," Clint said.

We walked to another smaller barn and entered through another coded lock. Inside there was a glass wall. On the other side of the wall was a large room with the pigs in it. There were black and white ones.

"The black pigs have the vaccine; the white ones do not."

"Some of the white ones seem sick," I said.

"They have the virus and no vaccine. They have all been infected for about eight hours. Pigs are actually more resilient than humans. We will check on them tomorrow, but so far it looks good." We walked back to the house.

"Can I get you anything," asked Clint.

"I'm fine thanks, a glass of water would be good. I did bring a nice bottle of wine." As he was getting things, he called Ward and told him to come down.

We sat on the deck in the evening glow of the setting sun, got caught up on recent events, and then opened the wine. It came from a friend of mine's vineyard – a Pinot Noir from Garnier wines. I'd been saving it, a vintage 2022. He said it was a very good year. Ward came in just as we were pouring the first glass.

"Looks like I arrived at the right time," he said.

"Indeed," Clint said, pulling out the syringe with the vaccine.

"Really? I arrive, and the first thing to take place is a shot?"

"Followed by some good wine," said Clint. "Where do you want it?"

"The wine in my mouth, the shot in my ass." We all laughed.

When we settled in Clint asked, "You said on the phone you wanted some more information, what about?"

"Michael said you would teach me more about amino acids and their import in the research you are doing."

"Yes, that is probably more in my purview than in his. I wouldn't want you to get the wrong information. Do you have that sheet I gave you on amino acids and DNA?"

"Yes." I dug it out of my bag.

"Here is another list of what those amino acids are responsible for. As you can see, they are very important."

1. Phenylalanine: Phenylalanine is a precursor for tyrosine, dopamine, epinephrine, and norepinephrine, which are neurotransmitters. It plays an important role in the structure and function of proteins and enzymes and the production of other amino acids.
2. Valine: Valine helps stimulate muscle growth and regeneration and is involved in energy production.
3. Threonine: Threonine is part of structural proteins such as collagen and elastin, which are important components of the skin and connective tissue. It also helps in fat metabolism and immune function.
4. Tryptophan: Tryptophan is often tied to drowsiness but has many other functions. It's needed to maintain proper nitrogen balance and is a precursor

to serotonin, a neurotransmitter that helps regulate your appetite, sleep, and mood.

5. Methionine: Methionine aids in metabolism and detoxification. It's also necessary for tissue growth and the absorption of zinc and selenium.

6. Leucine: Leucine is critical for protein synthesis and muscle repair. It also helps regulate blood sugar regulation, stimulation of wound healing, and production of growth hormones are other important attributes.

7. Isoleucine: isoleucine is involved in muscle metabolism and is concentrated in muscle tissue. It's also important for immune function, hemoglobin production, and energy regulation.

8. Lysine: Lysine assists in the synthesis of proteins, the production of collagen, elastin, hormones, and enzymes, and the absorption of calcium. As with others, it's part of immune function.

9. Histidine: Histidine helps in the production histamine, a neurotransmitter. This is important for immune response, digestion, sexual function, and sleep-wake cycles. It's critical for maintaining the myelin sheath, a protective barrier that surrounds your nerve cells.

"And that is only the big nine. This wine is excellent, by the way."

"Thanks."

"As you know, amino acids are the building blocks of proteins, which are the machines that do everything. In deep genetics, you have three choices. You can find or create the nucleotides that are what make DNA, you can create and piece together a very long string of amino acids, which is what DNA essentially does, or you can play with the DNA structure. Currently, we are working on all three levels. Working with DNA seems to be the best way at present because of advanced CRISPR technologies. I won't go into details, but what Simon can do is like comparing a floppy disk to a 64 GB thumb drive."

"Proteins are made of amino acids. How the different amino acids are strung together determines how they fold. What we have done is to change DNA and RNA strands to make them create the proteins we want. Those proteins, in turn, function as agents of destruction in the body. Think of this as an algebraic equation, albeit much more complex. We know that 3 x 5 = 15. If I take out two numbers and replace them with variables, it's a problem, unless you don't care what the solution is. If I give you 3 x ? = 15, you can figure out quickly what the ? is to solve the equation. If you didn't know algebra, you'd have to do trial and error. Keep trying numbers until one worked. To a large degree, that is what genetics has been about. We make guesses and then test the guesses to see if we get the right answer."

"Your machine takes some of the guessing game out of the equation," I said.

"Yes, about 70% of the guessing and getting better. That is up from 30% before Simon, Hypatia, and Patricia.

"How?"

"I'm sure you are aware of the Rubik's Cube?"

"Yes."

"How is it that some people can figure it out in minutes or complete it in seconds, while others never get it? Are those people stupid? I know PhD. scientists that never figure it out."

"I've never understood that myself."

"In the end, I believe it's their DNA and how their brain is wired; both are mostly out of their control. A 4x4x4 Rubik's Cube has 7.4 quattuordecillion or 10^{45} possible combinations. I believe everyone has a special gift or talent. Most of us never find them. We look at musicians like Mozart or Eric Clapton, writers like J.R. Tolkien and Isabel Allende, geniuses like Edison, Carver, Curie, or Einstein, or generalists like Benjamin Franklin, and others in every ethnicity on the planet, and believe they are unique. They are unique in that they found their talent. Everyone has one. The challenge for humanity is to unlock that potential rather than to imprison it. The answer is both a matter of nature and nurture. There are no commonalities among these types of people."

"Would you include yourself?" I asked.

"I include myself in the group that has a talent. I do not see myself as someone who has arrived at the potential of that talent."

"Do you believe you will?"

"I have hope."

"You have just created a vaccine that will cure the most destructive force the world has ever known. Isn't that enough?"

"No."

"You are a bit hard on yourself. You need therapy."

"Luly tells me that all the time. I am afraid a therapist would help me slow down and relax. If I did that, there would be no cure for malaria. People who discover their talent and love pursuing it and using it for the betterment of others, thrive. Most people don't want to thrive, and sadly, there are a great many who literally struggle to survive. I believe there are answers to those issues. I think Mary, Michael, and I are on the verge of helping with a few of them."

"For what it's worth, I do too."

"Back to amino acids and the Rubik's Cube. The ML and DL machines would methodically go through every combination. DL machines might find some patterns that would speed things up. What Michael has done is to tell the machine the endgame, give it an algorithm to research whatever it wants, other than the answer, and let it learn. I'm sure he talked to you about unsupervised learning."

"Yes."

"Ultimately, the Rubik's Cube is a mathematical puzzle. There are ways with advanced math you can figure it out. The math is easier to get to if you have the answer first."

"Just like DNA," I said. "I will say that I figured out the 3x3x3 cube, but it was not because of 'talent.' It was my grit, tenacity, and stubbornness that helped me. I did it once, I have no need or desire to do it again."

"You got it."

"Why don't more mutations stick in humans?"

42

"The ghost in the machine. I will be brief here. Let's say that you have 37 trillion cells in your body. Each of those cells is dividing to make new cells. When the DNA and RNA are dividing and multiplying, they are creating mutations. We are getting hundreds of billions of mutations per day," Clint said.

"Why are we alive?"

"The human body is miraculous. There are checks and balances to keep the mutations at bay. Some cells, like sperm cells, have far fewer mutations. We don't know how or why. Sperm and egg cells are the only ones that pass information onto the offspring. When you do things like smoke, you increase the number and severity of the mutation. Our immune systems take care of a lot of them. Some, like fast-growing cancer mutations, take over. We still don't understand that, but you have certainly heard of immunotherapy."

"Yes."

"The more we learn about the immune system, the more we learn how to help it. Your immune system has five parts, depending on how you look at things. Your skin is the first layer of defense. It keeps things from getting into your body – unless you have a cut or the infection is airborne. Remember, in Covid-19 that they had us wear masks and wash our face and hands?" asked Clint.

"Yes."

"Viruses exist outside the body on surfaces for a short time. If you wear the mask, you take away inhalation. But if you touch something with the virus on it, then rub your nose or mouth with your hand, you can get it. The chances are smaller, but still there."

"Then we have the lymphatic system. There are several parts to it. The lymph nodes gather lymph fluid and act as filters. They store cells that destroy cells that don't belong in your body. Your bone marrow creates white blood cells that fight off infections and the germs that can cause them. Here is how amazing the body is. There are cells called Basophils and Mast Cells that live in tissues. They look for antigens, which are markers that a cell is present that shouldn't be. When they find the antigens, they release histamine that attracts immune cells and antibodies to the site. When blood flows to that area, it creates inflammation, that helps keep the bad bugs from spreading."

"Then you have lymphocytes. These are white blood cells. There are T cells and B cells that I'm sure you've heard about. When something invades the system, or a mutation occurs that looks like an invasion, those cells help build an acquired immunity, so you don't get that again. They are not always successful, but more often than not, are. Some viruses seem to be immune to acquired immunity. They keep coming back, even though a person fought off the infection. The B cells make antibodies to kill invasive organisms. They also create antigens. T cells move from the bone marrow to an organ called the Thymus. This is like a nursery where they grow and develop and learn to differentiate antigens. They also learn about your body, so it doesn't attack the wrong thing. When mature, they live in your spleen, tonsils, adenoids, and intestine. They go out and destroy the bad cells. Some of them die, some stay alive as memory cells, remembering that particular bug, and ready to attack the next time they arrive."

"Remember our friends, the proteins? There is part of the immune system called the Complement System. This is thirty proteins working together to kill or mark germs."

"Sometimes, the infection is so large the body doesn't have enough to counteract it. Other times, the infection fools the immune system, like with malaria and HIV. The reaction is not strong or quick enough to fight the onslaught before it overwhelms the immune system. That is why we go to the source of the initial malarial infection and fight it there. It's like having laser beams at your front door, to zap a thief before they enter the house."

I'd been watching Ward as he listened to this. I could tell he was fascinated but clueless. His mind was churning, and he wanted to ask questions. Fortunately, he knew he was in grade school, and I was barely in middle school. He kept his mouth shut.

"Ward, are you getting all this?" I asked.

"This is old hat for me, didn't learn anything new."

We all laughed and clinked our glasses.

"I think I have had enough for tonight. You can all stay up as late as you want and discuss the depths of genetics."

"I would love to share my wisdom with the doctor, but I too will hit the hay," I said.

I was heading to bed. Ward said he'd make breakfast. My phone rang.

"Hello?"

"It's Linda." She sounded tense.

"Hi, are you OK?"

"They have me."

"Who?"

Another voice came on - a male.

43

"That doesn't matter. We have her, and her lovely life will be short if you don't do what we say."

I clicked my fingers at Ward. I made hand motions to call the helicopter. I said Linda a couple of times, and he knew what was happening.

"What do you want?"

"We want the software."

"You know I don't have it."

"You know who does, and you are protecting him. Better than you are protecting your girlfriend."

I wanted to say some bad things but knew that would not help.

"How do I reach you? I need to find a way of making this happen."

"It's simple, you call the software guy and have him send it to us, I will give you the email address. Oh, wait, he can trace that with his software. He will put it on a hard drive. He has six hours to do that. When I call back, that had better be complete, and then we will tell you how to make the transfer for your hot girlfriend."

"I need to talk with her."

"You have thirty seconds. No funny business."

"What is happening?" Linda said, clearly scared.

"I'm sorry. I'm working on getting you out of this mess. Did you take your medicine?" There was a pause.

"Yes."

"Good. You should think back to my second book, the last chapter, that will keep you distracted."

"Ummm, OK."

"Enough," said the captor. Six hours." The phone died.

"Damn, how could I be that stupid."

"Tom, you can't protect everyone you know," said Clint.

"I asked her if she wanted protection, she said no. I shouldn't have listened."

"You can beat yourself up later, what is the plan?" asked Ward.

"The chopper is on its way?"

"ETA one hour twenty."

"I pray she understood me," I commented.

I opened my computer. When the last book was published, I knew I had enemies, and she'd be an easy target. I gave her a tracking device she could take as a pill that would send a signal for 48 hours. I was waiting for the computer to try to connect to it. Michael's phone rang.

"Michael, I was just about to call you," I said.

"They have my son. I got a call. Someone has my son. He works for the CIA."

"I know. What do they want?" I said.

"You know they have him?" Michael asked.

"No, that he works for the CIA. What do they want?"

"My program, of course."

"What is the timetable?"

"In five hours, they will call. If I agree to their terms, they will let me know how to make the transfer," said Michael.

"What happened to his protection?"

"I have no idea." The call was on speaker. Ward got on the phone to the person watching his son. There was no answer.

"Someone has kidnapped Linda, my literary agent. They want the same thing in six hours."

"Perhaps I should make it public like Clint did," Michael said.

"If you do that, we have lost our leverage to get them back, and they are both as good as dead."

"I get your point. What is the plan?"

"I am going to make a couple of calls. I will call you back. Please do nothing. I assume you had Patricia trace the call?" I replied.

"She sensed the tension in my voice and tried, but all she could do was say it was within 100 miles of D.C. That is where my son lives."

"OK, hang tight."

I hung up and called the CIA. "Beth Wilson, please." I started to pace.

"Who is this?"

"Tom Armstrong."

"She's not in the office this late."

"You also know who I am. If I am not connected to her within three minutes, the world will know that you kidnapped one of your own spies. I will not sleep till I have brought the organization down. Am I clear?"

"Let me check."

Sometimes making threats didn't work. Sometimes they did.

"Tom?"

"Beth." I'd met Beth years before she became the Director of the CIA. She worked for the State Department as an analyst on China. She was brilliant. She was hired by the CIA and was indispensable in helping our government ferret out spies and create situations that put them in a corner. She was trusted by everyone and was apolitical.

"What is this about? A kidnapping?"

"I will not go into details. One of your operatives, Ben Longstreet, has been either detained or kidnapped."

"I know nothing about this."

"I hope that is true. If I find out it's not, the CIA will be my full-time job until it's gone."

"Don't make threats."

"Or what, you will kidnap or kill me?"

"We don't do that."

"Don't make me sick, of course you do, that is what you do, everyone knows it. But one of your own, that would be pathetic."

"How do I reach you? I will look into this, I promise. If we did it, it's a rogue operation. What do they want?" Beth asked.

"They want the software his father has created."

"We all want that, but I would not kidnap someone to get it. Perhaps if they were Russian, but not an American. The list is quite long of those who want that software."

"We know," I stated. "What does Ben do for you?"

"I'm looking him up now. I don't know who he is. Here. He's an analyst."

"I know you make a living lying. Analysts are not out of the country a lot, try again."

"You know I can't divulge what people do here."

"I want him found and on the phone with his father in the next four hours. Wherever he was two hours ago, I want someone sent. We had someone protecting him, and they are not responding."

"I will do my best."

"Pull out all the stops on this one, or I will." I hung up, probably not a wise thing to do with the Director of the CIA. I didn't care, but I did wonder how far I would have gotten if I had no clout. Not far.

"Ward, call everyone and tell them what is happening and to either be extra cautious, move those they are protecting, or both," I shouted.

"Tom, take a deep breath, yelling at us doesn't help."

"I'm sorry, you are right."

Clint was pacing.

The computer came on, and the blue dot that was Linda showed up in Soho, an area of New York City that had lots of warehouses. Many had been turned into lofts.

"Michael, I am sending you an address. Tell me everything you can about the building. Send it to my secure email. We are working on Ben. This is not his location, so don't have Patricia do anything – yet," I said.

"Clint, you will stay here. I want you planted by your security monitors. If you have the slightest concern, you call, do you understand?" I asked.

"Yes."

"Here is the address, Ward. I want two people near this address within the hour. Tell them not to approach the building."

"Got it."

"You seem quite calm for all that is going on," said Clint.

"The more intense the situation, the more focused and ruthless I become. I have never done well being put in a corner."

"I've noticed – gratefully."

"I hope we are as lucky with Linda and Ben as we were with you."

His phone rang. It was Beth. "We found your security. He's been shot but is OK. We are taking him to the hospital. He said three people in masks took Ben as he was getting out of his car. The family is fine. These are very smart people. They made him strip and get into new clothes. Unless he has a tracking device inside his body, we don't know where he is."

"Damn."

"Tom, I want these people as bad as you do."

"I doubt that. If it was someone in the government, who would the likely suspects be?"

"There aren't that many options of who could order this. You saw them at the Senate hearing. John Spencer, General Fleming, and I are the most likely. I can't imagine they would do this, especially Fleming; he's not interested in AI work."

"John has already stated he wants it bad."

"Tom, be careful. We all want it bad, but this is off the map."

"You use your resources, and I will use mine. Let me know what you find out."

"You ...," she started to say. I hung up before I could tell her I probably would not be giving her the information I had.

"Ward, tell your people to pick us up at the Hudson River Park at the west end of Warren St." He nodded. I paced. When things needed to get done, I was not good at waiting.

"What do you have?" I asked Michael when he called.

"You won't like it."

"Just tell me."

"It was purchased a year ago by Homeland Security and is used as an armory for potential terrorist attacks on New York."

"That narrows it down. Send me the layout, and can you look at the security and see what you can do?" I asked.

"On its way."

They hung up.

"Ward, it's the government. The building is an armory. The good news is, other than the security on the outside of the building. There won't be many people inside. I'm sure they are confident that with all the weapons, they are safe. They are also confident we have no idea where they are."

"Let's hope that is true."

I had to make a decision about what to tell Beth. She'd given me her direct line, trusting I would not abuse it in the future.

"Hi. I have some bad news," I said.

"He's dead?" responded Beth.

"Not that bad. I still don't know where he is, but I can give you some help, I think. You can't ask me how I know this, but you know me well enough to know I don't say things without having ample evidence they are true."

"I would agree."

"I think the operation is going through John Spencer. They have not only taken Ben, but also a person of significance to me. I suspect Ben is at some holding facility of Homeland Security. Do not call John and tell him you know this yet. Use your resources to see what you can find out about his location. We are headed to Soho."

"I will do what I can. I had a feeling John was going down the wrong road."

"I'm glad it was not you."

"Thanks."

I heard the chopper land. Ward and I ran out and climbed on board.

44

"Head to Hudson River Park, on the East Side," said Ward.

"That isn't an official landing zone," the pilot said.

"I know." He shrugged his shoulders.

"We will be going in very low over the water so as not to attract attention," the pilot said.

"Don't land, just hover, we will jump out, and you take off," I said.

"Not a problem."

I called Michael. "Michael, I want you to do two things."

"Name them."

"First, see if you can track any of John Spencer's phone calls over the past seven hours."

"Way ahead of you. Unfortunately, I think he is almost as smart as you. I think he used a throwaway phone like you do."

"Damn."

"He is not the brightest bulb in the house. He bought two phones with his own credit card. I traced the billing. They are required by law to include the serial number of the phone in the transaction. The serial number leads to the phone number. He's made five calls from that number. The latest is to the address I am texting you. I looked it up. It's a safe house of the CIA. Are you sure about Beth?"

"As sure as I can be. I can set up a rescue, or we can trust Beth and let her do it."

"Your call," Michael said.

"I'm going to trust Beth on this one. She knows the location and can get the necessary resources on the ground in minutes. It would take me hours."

"Then do that. I will send you a link. I will insist she patches any cameras they will be using through to me."

"Fair enough. I will need you to stop cameras and alarms at the armory building when I tell you."

"That shouldn't be a problem. We are already hacking into the cameras. They should be up momentarily so I can tell you where she is."

I called Beth.

"Good news and bad news. I am going out on a limb here."

"What?"

"We have located Ben."

"Where?"

"At a CIA safehouse outside D.C."

"That bastard."

"I agree. Right now, our concern is Ben. We will deal with John later. I am assuming you are still not in on this."

"I am not."

"Michael and I have agreed to let you run the rescue. Can you do it?" I asked.

"As you said, it is what we do."

"Here is the caveat Michael insists on. He wants the cameras the agents will be wearing to be patched through to the link I will send to your encrypted email. If those cameras go off, I won't comment on what he might do."

"Fair enough. Please tell him I would appreciate it if he did not record what is seen," responded Beth.

"That is fair as well, and I'm sure he will agree." I gave her the address and sent her the link. I smiled. You never know when the encrypted email account and phone number of the Director of the CIA might come in handy. I wouldn't abuse the privilege. "Please give me updates."

This time, she hung up. It was almost 1 am when we descended to a heliport on Staten Island.

"Why are we stopping here?" I asked.

"We aren't. I am letting radar see us landing. From here on, I will fly below the radar."

It took us five minutes to get to the park.

"Do you want me to pick you up?"

"Go park at the heliport nearby. We will meet you there in one hour if all goes according to plan. You will take us back to the farm where you picked us up." We jumped out with our bags.

The van was waiting with a guy and a gal in it. They introduced themselves as Liz and Jim. I had my laptop, and Michael had patched the cameras through. Linda was in an office at the back of the first floor. There were cameras on the front and back of the building, one looking at the large supply of weapons, and one on the roof." We parked the van a block away and got the team together as we watched the monitors. We saw three people. One was sitting in an office by the front door. There was probably someone there 24/7. Two others seemed to walk back and forth from the back of the building where Linda was to the front.

"OK, I will tell Michael to kill the cameras. When he does, we have thirty seconds to blow the doors and secure the building. The cameras will then come back on. There are plenty of blind spots between the cameras. Ward and I will go through the back, you two through the front. I would prefer them alive. If there is a problem, shoot them. Either way, you have thirty seconds to get them off the camera. At two minutes after entry, Michael will kill the cameras for another forty-five seconds. Take the person or body out the door. The van will pick you up. If they are dead and there are people on the street, make them look like you are helping a drunk. Everyone understand?"

They all nodded. I called Michael, and we synchronized watches. We started to walk around the block toward the doors.

"Wait," shouted Michael.

"What?" I asked.

"Something is wrong."

"What is wrong?" chimed in Ward.

45

"Patricia caught it. There are four people in the building. They are all dressed the same, but she noticed the different ways they walk. When the two go back and forth to the back, one ducks into the stairwell that is in one of the holes of the camera. I don't think she is downstairs. Why would they do that?"

I pushed my brain – think, think, think. Then it hit me.

"Damn. I got it. I think. I have written several thrillers. I told Linda to think about the last chapter of the second book. Everything we are about to do is premised on that chapter. I think she thought I was talking about the second book she was the agent for, not my second book. What are the odds they gave her a truth serum, and she told them what she thinks was going to happen? I think they are ready for us. They will know we are coming when the cameras go out. We have two choices. We can either go in with cameras rolling and surprise them, or have Michael do a series of on and offs until they relax. We still don't know where she is."

"The building has five stories. According to the layout I have, there are offices on the second floor, but the cameras aren't working. There is nothing on the rest of the floors," said Michael.

"OK, someone use the infrared scanner and see what you can see by walking around the building. Michael, start turning the cameras and lights off and on. Do so in thirty-second intervals at first. Tell us when, we don't want to be in camera view when they are on."

"Ready?"

Ward put the goggles on. "Go."

"Ready set go."

The building went dark. Ward dashed down one street looking up. Nothing. He stayed on the back corner, waiting.

A minute later, the lights went out. He ran across the back of the building. Nothing. On the next side, he saw two heat signatures near a window. It looked like one was standing, and one was seated. Finally, he checked the front of the building. Nothing.

"I believe they are on the north side of the building in the middle. It's the only room with a light on when the electricity is on."

"All right. Change in plans. Silencers on guns. They will be wearing Kevlar vests since they seem to know we are coming. Michael, keep the on-off going, but change the pattern."

"It takes forty-five seconds for the backup batteries to kick and reboot the system. I have let them reboot each time so they can check the cameras."

"Same arrangement as before. Synchronize watches again. In three minutes, Michael will turn on the cameras. We wait one minute five seconds for the cameras to reload and give them a glimpse that things are OK. Then we go in. Ward and I will go in the back entrance, closest to the stairs. While I would like one of them alive, I don't really care. If you have time and are safe, shoot them in the leg or shoulder. Otherwise, in the face. Do not hesitate."

Everyone nodded, got their weapons out, and headed for the corners of the building.

"Three, two, one." The lights came on. Tick tock. Ward had brought several explosives that would blow the doors in. Once on the door, we had two seconds to run.

"Now." He ran and put the explosives on the door and came back to me. The explosion was severe, and there wasn't much of the door left.

At the front of the building, the explosion almost knocked the guard in the office off his chair. He was not holding his gun. They shot him in the shoulder and the leg. Our person ran in and cuffed him as he screamed in pain.

Ward and I ran in the back of the building. The second guard was in the middle of the building, not sure which way to turn since the front door had imploded as well. He must have been a recent recruit. He dropped his weapon and put his hands over his head.

I learn the hard way at times. I went to put cuffs on him. He turned and clocked me on the head. I fell to the ground, embarrassed and sore, but conscious. A second later, he dropped to the floor with a bullet in the head. I stood up, looking at our helpers.

"I'm sorry you had to do that. My bad."

"Don't worry. The other guard is cuffed."

"You two stay down here. Let us know if you sense any issues." He nodded.

I joined Ward at the foot of the stairs. He'd shot the other guard in the leg. He was sitting on the floor, moaning. Whoever was up there knew we were coming and had probably called in backup. I gave us ten minutes max.

No one in the hallway.

"We know you are in there. If you want to come out alive, you will throw out your gun," I shouted.

"I don't think so. I have a gun to your girlfriend's head."

"Tom, don't do this," Linda said with fear in her voice.

"Hi, Linda, just remember." I was thinking of the book she was thinking of now.

"I remember everything."

"I know you do."

"All right, you two, shut up. I don't know who you are, but you have ten seconds to throw down your weapon and come in the room, or she dies."

"This is Tom, the person you called. Here is my weapon. I will come around the door with my hands raised. I will turn around, so you see I have no other weapons."

"Where are the others?"

"Tending your friends downstairs. I figured we'd only need a couple of us. I didn't expect four of you. There is no way out of here for you."

"I think I have a plan."

I leaned over and whispered in Ward's ear. He nodded.

I threw the gun through the door and stepped into their view. He was standing behind her, using Linda as a shield. He was a little taller than Linda and broader in the shoulders.

"I know we can work this out," I said.

"There is only one way this works out with both of you alive. Get me the software I asked for. My reinforcements will be here soon, so make it quick if you want to live."

"I'm going to reach for my phone and call the person with it, is that OK?"

"Yes, slowly, no tricks, or you are gone."

"No tricks." I pulled out the phone with a gun pointed at me. I put it on speaker.

"Michael, how is the software coming?"

"Tom, I'm having second thoughts. I mean, this software will potentially impact billions of lives. We are talking about one here."

"Enough, Michael, just do it."

"Yea, stupid, do what he tells you to do. I will kill them and come for you next," said the captor.

"Hey buddy, you might tone it down a notch."

"I will do what I want," he said, clearly agitated and distracted by the conversation.

"Listen…" I said. With that one word, Linda leaned to the right. Ward came around the corner and shot the guard in the face. He fell, but not before he let off a round that caught me in the chest. I fell to the floor.

Ward ran over, put another bullet in the guard to ensure he wasn't going to cause issues. Then he undid Linda.

She ran to me. "Oh my God, you are hit. Call an ambulance."

"Too late for that," said Ward.

I opened my eyes. "Do you want me to shoot you too?" I asked, smiling.

Ward looked at Linda. "He has a vest on, he's fine. He will just have a bruise that I'm sure you can help him with."

The others came up to check on us.

"We need to get out of here now. Jim, bring the van around to the back. Liz, bring the live one from the front to the back." They both ran.

A minute later, OK, maybe two, I was not moving that well. We piled into the van and left.

"Michael, we are all safe. Tell me when the reinforcements arrive." It was less than two minutes. And thanks for turning all the cameras off. Wouldn't want to be maskless on camera."

I turned to the one called Jonah. We got their names from the ID in their wallets. "Who do you work for?"

"The one you killed?"

"What's his name? He carried no ID."

"Wouldn't you like to know."

"Yes, I would, and I think I will soon."

"I don't think so."

The pilot was waiting, the helicopter on. We parked the van beyond the reach of cameras. I stopped by the payphone and made a call, then Linda and I hobbled to the ride. The pilot had covered the numbers so they couldn't identify it. There was no one there anyway.

"I can't take everyone," said the pilot. "Too much weight."

"No worries. Just taking five and the trip for one of them will be short. Jim, you and Liz take off, if that's OK with you. Ward will settle up with you in the next couple of days. Is that OK?" I said.

They nodded and left.

46

We took off. I yelled at the two guys since the noise was loud.

"One of you might live till tomorrow. We are going up to three thousand feet. The first one who tells me who you work for lives, the other learns to fly, and I hope you can swim."

"His name was Mick," said the guard in the office.

"What is your name?"

"Victor."

"Is Mick the boss?"

"He hired us."

"Are you freelancers?"

"Yes."

"Who did Mick work for?"

"I don't know."

"OK, I guess you go first." I moved him to the door.

"Stop. I heard him say, 'Yes sir, Mr. Spencer,' during one of the conversations he had."

"That helps."

"Does that sound right to you?" I asked Jonah, the other person.

"He's full of crap."

"Who do you say he talked to, and what is his name?"

"Santa Claus."

"I hope you enjoy your flight." I kicked him out of the plane, screaming. While in the chopper, I'd attached weights to each of them. I didn't think he would ever surface. Hopefully, he wouldn't be missed.

"Now, Victor, I need to know everything or you join your friend."

"I don't know much. Mick called me and said he had a job. I asked what it was and who it was for. He said the government and it was an easy $20,000. One night's work."

"Was it worth it?" asked Ward.

"No."

Linda reached over and slapped him hard. "Personally, I wish he'd throw you out of the plane. Your mother did a bad job raising you," she said.

"Good girl," I said, giving her a hug. "Ouch. Don't squeeze so hard."

Ward laughed. "You are such a wimp."

"I know."

"Mick made a call after you called and talked with Linda. He asked her some questions, slapped her a couple of times, and then injected her with something. Probably truth serum. After she was done talking, he said there was a change in plans. That's when we moved her upstairs. He told us about the plan for you to kill the lights and enter. When that didn't happen, we weren't sure what was going on. Then you came in," said Victor.

"Linda, you almost got us killed."

"How?"

"When I said the second book, I meant my second book, not the second book you were the agent on. Fortunately, Michael and Patricia knew something was off. We changed plans, except for the very end. I'm glad you leaned when I said the word listen. If you hadn't…"

"You'd be dead," said Ward.

The chopper was a bit crowded, and she was sitting next to me. It was dark inside. She reached down and grabbed me, bit my ear gently, and whispered loudly, "I want you now." Adrenaline rushes and being saved had that impact.

I looked at her, and my feelings echoed hers. I had no doubt another dinner was in the near future.

I called Beth. "Beth, how are things going on your end? Our end was successful and is a wrap."

"We just secured the building, and Ben is fine."

"You have a cleanup on aisle four." I gave her the address and outlined what happened. "The ringleader, named Mick, was on a phone call and was overheard saying 'Mr. Spencer,' to his boss. The reinforcements were on their way. They may have cleaned up already."

"I doubt it. We knew the area you were going to. We have people arriving now."

"We have one living person who was part of the kidnapping. What do we do with him?"

"Where are you heading?"

"We can drop him off at Langley if you want."

"I'd prefer Annapolis if you don't mind," she said.

"Fine with us. Will you let them know not to shoot us down?"

"ETA?"

"30."

"They will be waiting."

"What about Mr. Spencer."

"I have a call into the President."

"It's 2:00 am."

"He's a light sleeper, and Mr. Spencer is a pretty important person. The President would not be pleased if I waited."

"Tell the President that I will be less than happy if an excuse if made for his treason, and he walks."

"I will certainly let him know you said that," I smiled.

"Has Ben talked to his dad?"

"They are speaking now."

"Thanks for doing this. I owe you. My apologies if I came on strong."

"I understand and would have done the same. But I will keep that chit you owe me in my drawer," said Beth.

"Not a get out of jail free card."

She laughed. "Tom, I understand that."

"I'm starting to like you."

"Be careful."

"That is why I am still alive," I said.

"You know where to find me."

"I won't abuse that privilege."

"I know. If I didn't trust you, I'd change the number," said Beth.

"I'm sure you could do that."

"Good night."

"Night, Director."

We landed at the Naval Academy. A van was waiting to host our kidnapper. I trusted I would never see his face again in public. If I did, he was dead. The pilot had radioed ahead, and a fuel truck was waiting to fill the helicopter.

"I figured they could spare a little fuel," said the pilot.

"The least they could do," I added.

Our pilot landed on Clint's property at 4 am. Clint was up cooking breakfast.

47

"Hello, Linda, I assume you are hungry?" asked Clint.

"Yes, the flight offered no food," Linda replied.

"I was a bit preoccupied," I said.

"I know. You are forgiven." She kissed me on the cheek.

Clint walked to the pilot. "I hope you will stay for breakfast."

"I would enjoy that."

We all sat at the large table on the deck. There was ambient lighting and a full moon. The temperature was warm enough, and the air was still.

"I didn't think I would ever savor peace and quiet as much as I do right now," said Linda.

"I agree, quite a night. Clint, this is fantastic. When you retire, you should open a restaurant. At least fill in down at the Bonefish," I said.

"I'd fly in for that," said the pilot.

"Having been kidnapped and rescued myself, I'd be interested in hearing about it. But I understand if you'd rather not," said Clint to Linda.

"I stayed late at work and was alone in the office. There was a loud knock at the front door. We don't have cameras to see who is there, and we are on the 10th floor of an office building. I opened it, and two men with masks grabbed me. I screamed. They put their hand over my mouth. I got enough control to point to the desk. They said if I screamed, they'd put me out. I nodded. That is when I told them about the medicine, the pill you told me to take. They let me take it. Then they must have put chloroform or something on my mouth. I passed out. The next thing I remember was coming around in a van. I had tape over my mouth, and my hands and feet were tied. We didn't travel very far. I don't think we did anyway; I was unconscious. They took me into the building. I was confident we didn't leave Manhattan. Nice of you to think of that and to think that I might be kidnapped."

"You said you didn't want protection, I offered," I said defensively.

"No need to get defensive. I was grateful I had that device. Is it still in there?"

"Yes, it works for about 48 hours, then comes out – naturally."

"Good, I don't like the idea of you knowing my every move."

"I know plenty of your moves."

Everyone laughed, and she socked me in the arm.

"After the phone call, they tried to get me to tell them what we'd said. I refused, and they gave me a drug. It made me woozy. I don't remember what I said, probably the wrong thing. Then they took me upstairs. When the lights started to go on and off, they were on full alert. Then, after a while, one of them said it was just the crazy electricity in Soho. They relaxed, and you came in. That was a good tip you gave me about the lights. Sorry about the mistake with the books. I knew I was going to be rescued."

"What is she talking about?" asked Clint.

"There was just one problem with the light thing. She thought I was talking about the second book I wrote when she was my agent. I wrote a few before that. I meant my second book."

"Ooops."

"I'm glad we figured that out. Well, Patricia did," I said.

"Who is she?" asked Linda with a bit of a jealous tone.

"More on that later, she's a computer."

"OK."

"When we figured out which book you'd read, we had to change plans. The word that was the clue for you to lean to your right was different in the correct choice, and you would have leaned the other way."

"And been dead," said Ward.

"I appreciate the heads-up."

"You write about a lot of kidnappings, I take it?" asked Clint.

"They do happen, as you well know. In one of the books, it takes longer to figure things out. Patricia and other technologies have probably saved both your lives."

"We appreciate that reality."

"The four of us were in and out in under five minutes. We left two dead, one accidentally fell from the helicopter over Delaware Bay, and one is in the hands of the CIA. It turns out John Spencer, Director of the NSA, was the ringleader."

"Will things calm down now?" asked Clint.

"I hope so, but don't hold your breath."

The phone rang.

"Hello Michael, did you have a good talk with Ben?" I put it on the speaker.

"Yes, thank you for making that happen," Michael said.

"The CIA is not all bad; we owe Beth one."

"Duly noted, as has Patricia."

"Is Ben OK?"

"He seems unshaken. Let's just say I don't think he's an analyst. He did say he was rethinking his line of work in light of this," said Micheal.

"Wise decision," I commented. "Where are you staying at this point?"

"I will head to the other place today."

"Good. Sonya is going as well?"

"Yes. She has been staying in the Co-op for the last couple of days with the added security we have outside. She's quite smart. I like her."

"Don't get any ideas. I don't need her compromised at this point," said Ward.

"She is all business."

"I know how that goes," I said, looking at Linda.

"Ah geez, keep it together," Ward said.

"Why?" she asked.

We all laughed again. For having just returned from being kidnapped and killing people, we were doing pretty well or were well into denial. I didn't really care which.

After breakfast, Linda said she wanted to go for a walk on the beach. I said goodbye to the pilot, who I assumed wouldn't be staying long. Dawn was breaking.

We were barely out of sight of the house when Linda almost ripped her and my clothes off. She grabbed hers and ran to the beach, laughing loudly. Of course, I chased her.

Splashing in the water, we hugged each other. She looked me in the eyes.

"Thanks for rescuing me, my knight in shining armor."

"Your welcome fair maiden?"

"How can I repay you?"

"I think you know."

We were standing on the edge of where the waves came in. She pushed me onto the beach and straddled me.

"Ouch, careful where you push. I have bruises where I took a bullet for you," I said.

She winced. "I'm sorry, let me kiss you and make it better."

"That will work."

I loved the feel of entering her tightness. She leaned over, her breasts pressing against me, her tongue feeling like it was going down my throat. I felt her rise and fall on me. I wrapped my arms around her as we made love.

The helicopter flew about fifty feet over us. He blew a horn. She sat up and waved, giving him a nice view. We didn't care; he'd helped save her life.

When we were done, she got off. I sat behind her, looking at the ocean. One of my hands was caressing her breast. The other moved south and filled her. She turned her head and kissed me as she orgasmed. We collapsed in each other's arms as the waves washed over us.

The water was not that warm, and we soon rose, put our clothes back on, and headed to the house.

"I wonder what those smiles mean?" said Ward.

"OK, OK, what next, let's focus."

A sound went off.

"Shit," said Clint.

"What?" I asked.

48

"That's an alarm, someone is coming," Clint responded.

We all went and looked at the computer. Three SUVs were heading to the house.

"Who could know we are here?" I asked.

"Don't do anything yet," said Ward.

We watched. The cars parked, and six military people got out with their weapons. Finally, the second SUV opened, and General Fleming got out.

"I should have known, but how did he know we were here?" Clint asked.

On the other side of the SUV, Susan got out.

"Damn. I knew she couldn't be trusted," he said.

The General and Susan came to the front door. The others surrounded the house. There was a knock at the door.

"Clint went and opened it.

"Hello Clint, I know others are here. I want everyone in the room, weapons on the ground, shirts off, except the lady. Susan will check her out."

Everyone did as he said.

We were all in a line against the patio window.

"What do you want, General?" Clint asked.

"You know full well what I want. I want the virus I know you have."

"I won't give it to you. You will use it against the world."

"Have you been asleep the last 60 years? It's them or us. The Russians and Chinese are waiting for a weakness. We must strike before they do."

"That is not the solution."

"I suppose you have a better way to ensure the world is safe from communists and crazy dictators?"

"I think so."

"I doubt it. Give me the formula."

"No. How did you find us?"

"You aren't as clever as you think. Because of national security, I had a conversation with the man that delivered your new machine. He let me know about the other one and where he'd delivered it. He was concerned you wouldn't like it if he did. I told him he could think about that in jail. He opened up."

"It just takes one mistake. I should have had him drop it off somewhere else and brought it here myself. Oh, well," Clint said.

"Yes, the things we don't think of. Fortunately, I have all the bases covered. If I have to kill all of you, I will. Then we will break into your computer and get the formula. You have the blessing of knowing your family, and that of Mary's will die a slow death."

"You are a monster."

"I am a savior. There is always collateral damage. I hope you are ready to die."

The General raised his gun, pointing it at Clint.

A shot rang out.

The General had a look of surprise on his face as he dropped his gun and grabbed his heart that now had a large hole in it. He stared at Susan.

"How could you?" he said.

"One of the easiest things I have ever done," she said.

He dropped dead on the floor.

"You just killed a General," said Linda.

"Should I take it back?" replied Susan.

"No."

"Good. The others will be coming in soon," said Susan.

"Do not go outside," said Clint, running to his console and pushing buttons. "When I say go, you will probably have ten seconds to get them before they regain balance and sanity." He flipped the switch.

Outside, a very loud piercing sound penetrated the air. All the soldiers grabbed their heads and fell to the ground. He let it go for ten seconds, then stopped it.

"Go."

Ward grabbed the General's gun and ran outside. He wanted to kill them but realized the General probably lied to them, and they were just following orders. He shot the two in the leg. Susan ran out front and shot the two on the ground, one in the shoulder and one in the gut. She collected their weapons and brought them inside.

Clint screamed.

"We have a problem."

I went to the console where he was.

"There is a beach landing. Who are they?" asked Clint.

"Not ours. Must be 30 of them. They aren't bashful. They are either Chinese or Koreans," said Ward.

"This can't be happening," said Clint, almost in shock.

49

Ward came in. "What's happening?"

"An amphibious assault. I think the Chinese."

"What are the odds?" I asked.

"Slim to zero, and yet it's happening."

"May I doctor? I can stall them for a few minutes," Ward said.

"Doctor? Ideas?" I asked.

"Follow me."

"Everyone follows the doctor. I will get this going and join you. Where do I go?" Ward asked.

"I will leave the door open to the barn, come in and close it behind you."

We all ran to the barn. Ward had assumed if there were an assault, it would come from the sea. He'd planted a series of mines around the beach. They were timed to go off every eight seconds for two minutes unless he stopped them.

"You will have to trust me," said Clint to the others as they entered the barn.

In the first room, he pushed the button and went through the second door. They crowded into that room as he opened the door to the Level 4 room. They were wearing no protective garb.

"Tom, check the monitors." He pointed to a console.

"I can see the mines going off. They are slowing them down."

Ward entered a few minutes later.

"Doors are sealed behind me."

"I think they will go to the house first, especially seeing the bodies. Wait till they are thirty feet from the house, then turn on the sound."

"We have another problem," Ward added.

"What?"

"There is a group of military trucks coming down the drive. They are ours."

"The General must have called in backup. They were probably waiting for a call from the General saying everything was OK. When they didn't get the call, they decided to come."

"Right now, they are the enemy. I don't think if we give ourselves up, having killed the General, they will be understanding."

"I agree. The Chinese will be their first battle, however."

Within a minute, gunfire started to erupt. There were more Chinese than Americans, but the Americans surprised them, taking out ten before they knew what was happening.

The Chinese from the beach landing were starting to arrive at the house. Soldiers from both sides had taken cover around the house, barn, and in the woods. There was probably a quarter acre of open space in front of the house – no man's land.

"Ward, you will see an icon with a microphone. That puts whatever you say out to speakers on the barn and house. Next to that is a translator. It will translate what you say into 20 languages."

"Before I start making threats, do we have to go out the way we came in?" asked Ward.

"No, I told you I was paranoid." He smiled.

The doctor was busy with the vats and with the enclosed box that held all the really bad bugs.

"What are you doing?" I asked him.

"We may not be coming back, so I need to take what I need for the rest of the exercise."

"Clint, you are taking Level 4 materials out of here? Into the open air?" asked Susan.

"Yes, there is no other way. They are very protected. I will offer each of you a vaccine if you want."

"For what?"

"The virus we have created."

"You told me that wasn't going to be ready for another few months," Susan said.

"I lied. I didn't fully trust you. Now I do, you shot a General."

"I appreciate the vote of confidence."

"How has this all been tested?" I asked.

"Ward, click on the cameras and look at the one labeled pig barn."

It came up on the large screen they could all see. Some pigs were lying down or dead, others were eating or roaming around, and a few were clearly struggling. The pictures were all in black and white. The ones on the ground were white, as were the ones struggling. The others were black.

"I painted them before we started. The black ones have the virus and the vaccine. The white ones have the virus and no vaccine. It was given to them two days ago. I would say the vaccine, at this time, is 100% effective."

"At this time," said Susan.

"We didn't have time to do longitudinal studies. We will. Hopefully, they won't be necessary."

"What is the plan?"

"One step at a time. Who wants the vaccine?"

They all got in a line. The doctor administered it to everyone but me since I'd already injected myself with it.

"If everything works, you should feel tired, and your eyes will water a bit in about six hours. If you have other symptoms, you need to let me know immediately. It does not mean it's not working; I just need to know," Clint said, with an admonishing voice.

Ward went back to the microphone. "Attention. You are on private property that is well posted. If you have a warrant, please walk to the center of the open space and show it. If you do not, leave the property now, or you will not leave alive. You have been warned. I do not care if you are the military." He then pushed the translation button and saw the Chinese soldiers listening. He didn't expect it to change their minds.

"OK, doctor, what is the plan?"

"I'm just about ready. Is everyone dressed for a walk in the woods?" Clint asked.

They all nodded in the affirmative.

"Ward, push camera button 17 and 18."

He did so, and two images came on the screen.

"Those are infrared of the woods. No one is there, that is good news," said Ward.

Outside, gunfire was still going. With all the cameras, they could see soldiers falling to the ground. Ward turned on the sound machine. Everyone near the house or barn grabbed their ears. He waited thirty seconds. Their ears would be ringing for days, and they would be incapacitated for several minutes.

"OK, I'm ready, let's go," said Clint. He pushed more buttons on a wall, pushed a table out of the way, and lifted a hatch in the floor.

50

"A tunnel, I should have known," said Ward.

"You have thought of everything Clint," I commented.

"I hope so."

"Susan, you first, then Tom, Ward, Linda, and me."

They all followed instructions.

"How far does this go?" I asked.

"Almost three hundred feet."

The tunnel was very narrow and a little over six feet in height. They'd each been given a flashlight.

"Does anyone get claustrophobic? Ward asked.

"After about thirty minutes, I start to react," I said.

"We won't be here that long," Clint said.

He turned on his tablet and checked what was going on topside. There were no people in the woods, and most of the soldiers on both sides seemed to be dead or incapacitated.

"What do you do about the mess at the house?" Clint asked.

"We have to think about that. Because of the Chinese involvement and our military going way out of bounds, I think either the FBI or the CIA. I would call Homeland Security, but I do not know if John Spencer acted alone," I said.

"I agree. You have a good connection with Beth at CIA, call her when we get out," said Ward.

"Make sure you let her know to clean things up and not poke around. Patricia is watching, and if she sees anything unsavory happening, she will blow the place to smithereens," said Clint.

"Susan, you on board with that?" I asked.

"Yes. I will have some explaining to do."

"We have your back. Do not see anyone without someone with you," Ward said.

"I've heard that before," I commented, smiling.

"As soon as we are away from here, everyone calls your families and makes sure they are OK. Ward, phone your team. How are we getting out of here? Our vehicle is at the farm?"

"On a raid like this, usually there is one person that stays with the vehicles. They will be within two hundred feet of the driveway. Ward and I will scout them out and take out the guard – alive or not, their choice," I said. Ward nodded in agreement.

"OK, here we are. Ward, twist the crank and push it up. There shouldn't be a problem. I go out once a month and open it from the outside and clear away any roots that might be overgrowing the hatch," Clint said.

Ward took a step up the ladder and pushed. The lid opened easily. He had no idea where we were, but he climbed out and helped the others, then closed the lid.

"I don't want them to find this, so everyone goes a different direction, then circle back and meet at that large tree over there," Clint said, pointing to a tree 150 feet away. We all started walking, trying to be careful not to flatten plants or leave a trail.

"The road is one hundred feet in that direction. You come out one hundred eighty feet from the driveway to your left. My guess is the military vehicles will either be right where you come out of the woods or across the road on an old dirt road about fifty feet closer to the drive," said Clint.

Ward and Susan took off. They used hand signals to tell each other what they were doing. There were no vehicles on the main road. They spotted the guard through the trees.

"OK, I will have my military hat on. I will approach him on the road. If he lets me approach, I will disarm him, and you come out. If he raises his weapon and looks or sounds like he will shoot, take him down," said Susan.

"Got it."

Five minutes later, Ward saw Susan approaching the guard.

"Soldier, I am Captain Susan Fellows. Put your weapon down. The mission has been compromised. I am here to aid in the clean-up."

"Captain Maam, I was told no one but General Fleming leaves without the General's permission."

She kept walking forward.

"Private. I was with the General when he was shot. He is dead. I barely escaped. He won't be calling."

"I have my orders."

"You have new orders, and if you don't want to get court-martialed, I encourage you to obey."

He hesitated but put his gun down. She approached and took it from him.

"Ward."

Ward came out from the woods.

"Check and find a car that is big enough for all of us."

"Including him."

"No, he will be staying."

Ward called the others and told them to come. He checked the other vehicles and picked a large SUV. He gathered all the keys and shot all the tires.

"I don't want them going anywhere."

"Why all the tires? Wouldn't one per car be enough?" asked Clint.

"They have spares, and this is the military. They would strip one car to make another work."

"Good point."

Ward disconnected the horn on the vehicle the soldier was now handcuffed to. We all climbed in the SUV and took off. I told Linda to drive while we made our phone calls, then Ward would take over.

"Beth, this is Tom Armstrong, sorry to call."

"Not another issue, I hope."

"Yes, a rather large and sensitive one. Again, I have more than enough evidence to prove what I am saying. I will give you an address. General Fleming is there, dead. He committed treason, and a soldier shot him. You will find another 20-30 American and Chinese soldiers on the property. They were trying to kidnap Clint. It's his farm. I know this is not your norm, but I don't trust the NSA, and with Chinese involvement, I wasn't sure if the CIA or FBI should be brought in. If you feel this is FBI material, call them. We have no idea who the soldiers may have contacted. I will tell you this, and you can relay to anyone approaching the house. Dr. Williams is a very paranoid person, with good reason. He has extreme security provisions, some of which disabled many of the soldiers on both sides. We are watching the property. If anyone, and I mean anyone, attempts to block the cameras or enter the barn, he will self-destruct the property. Those within a hundred yards will no longer be on the planet. Am I clear on this?"

"Very."

"Good."

"Right now, get those people's bodies off the property. When that is done, call me, and we will meet, and I will give you all you need to know."

"Fair enough."

"What about Spencer?" I asked.

"He was arrested less than an hour ago, trying to board the NSA jet."

"Don't tell me – he screamed innocence."

"Of course."

"I think he and the General were in this together."

"I never liked either of them."

"That is good news."

"Who is with you?"

"I will keep that to myself at this time. Have your people prepared. The General and the Chinese may have called reinforcements. Do this quickly, please."

"Thanks for the tip, people are already in the air."

"We see no movement on the ground, but be careful. You can land in front of the house, on the main road, or the beach."

"I will call you and let you know."

"I will know because I will see them. Make sure you have a lot of body bags. We only killed the General. We wounded four others. If there are Americans dead, it was the Chinese."

"Duly noted. Trucks are heading there from Philadelphia. I will be in touch."

"Thanks."

"Ward, any word from anyone?"

"Everyone is safe. Michael is moving to his other place tomorrow." The rest of them were off their calls.

"OK, everyone – breathe," I said.

"Where to?" asked Ward.

"I need to see Michael before I leave," said Clint.

"You can drop me off at my place; I need to clean up and check in with work. I haven't had a shower in days," said Linda."

"We will drop you off at the transit center in Edison, will that work? Are you sure you are OK? I mean, it's been a rough twelve hours for you," I said.

"I'm fine, or still in shock. The transit center would be fine. I suppose at some point I will need to let the authorities know what happened."

"Don't agree to see them without letting me know. Beth from the CIA is running interference for us on this one," I said.

"I will call you."

"Susan?"

"I'm not sure. I'm nervous about going back to Fort Detrick at this point."

"Come to Michael's," I said.

"I don't want to impose."

"He'd be upset if you didn't, I texted him already. He said, of course."

"I guess that's settled," she said, smiling.

"Get off at the next exit," I shouted.

"What's the problem," Ward asked.

"Trackers, we haven't checked the truck. The military tracks everything," I commented.

"Damn, my bad. I don't have a tracker finder, do you?"

"Yes. I brought mine. The military doesn't use advanced trackers on vehicles like this. I mean, where are they going?" I asked.

"True," said Susan.

We pulled off the road. I walked around the car but found nothing.

"Take the batteries out of your phones unless they are untraceable phones. I think Michael can make them safe at his house. Do not use them under any circumstance without checking with me. Understood?" I asked in a demanding voice.

Everyone did.

"Can anyone think of anything else we need to think about? We can't have anybody finding us at Michaels," I said.

"I understand you still may not trust me. I will be blindfolded the rest of the way if you'd like," said Susan.

I looked at my phone. "That won't be necessary," Michael says.

"How did he know?" asked Ward.

"Michael hacked into the Bluetooth in the SUV. I gave him the license number."

"Who is this guy?" Susan said.

"Exactly," said Clint. The pilot had radioed ahead, and a gasoline truck was waiting to fill the helicopter.

51

We arrived at Michaels two hours later. His giant log cabin sat on the edge of a lake. It contoured up the side of a steep hill, the entry to the house five stories above the lake, looking over the tops of most of the trees. There was actually a small floor above the entry. I guessed that was his hideout.

"Welcome, come in. I will give you a quick tour. You will need to get used to Patricia talking to you, especially when she suggests you not go into a particular area or room. I am not paranoid, of course, but Patricia is."

"Michael, I do not believe that is true," said a female voice that seemed to come from nowhere."

"You are right, Patricia. I am the paranoid one."

"You have good reason to be. There are people after you, or at least after me."

Every turn gave a new vista. His walls were covered with an eclectic array of art. Music seemed to ooze from the walls. As we got the tour, he pointed out people's rooms. We came to the living room that had a glass wall looking onto the lake. A twenty-foot tall stone fireplace was just off-center. He didn't want it interfering with the view.

"Can you swim in the lake?" asked Susan.

"The water is getting a bit chilly now, but be my guest. This used to be a camp. I bought the camp when it went under including all the land around the lake."

"This is spectacular Michael, nice job," I said.

"Thanks."

He made us drinks, and we sat on the deck, filling him in on all the details of the past two day's events. I was exhausted just thinking about it. While I'd seen the guitars at his New York place, I hadn't heard him play. That evening, he and Clint gave an improv performance I wished had been recorded. He let me know that Patricia recorded everything. The past hour's session was already in my email. The music, like Clints, was a fusion of classical, jazz, sacred, and rock. They acted like they'd played together for decades.

"Clint, thanks for that, I haven't had that much fun in a long time," Michael said.

"Me either, don't tell Luly," said Clint.

"I don't mean to get down to business, but what are the next steps?" Michael asked.

"I go to Morocco tomorrow, then the conference in six days, and D.C. in ten. The testing will be done by then," Clint said.

"Testing?" asked Susan.

"Yes, we are doing more testing of a few vaccines we are working on. This is how we did the malaria vaccine," said Clint.

"Michael, I hate to interrupt, but an emergency session of several committees of Congress has been called," said Patricia.

"Can you record any of that?"

"I am sorry, but there are no electronic devices allowed during those sessions."

"Damn," Michael said.

"They confiscate cell phones before they go into those sessions," I added.

"I suspect they are talking about us."

"No doubt."

"How do you think they will spin the attack on Clint's place?" Michael asked.

"To be honest, I don't think it will see daylight. No one knows about it, and if they broadcast the event falsely, they know I will spill the beans. At this point, they don't want to create an international incident. They will let the Chinese know they know, and there will be a price to pay. How the Chinese thought they could get away with this, I don't know," I said.

"Why, Clint?" Michael asked. "I understand why they would come after me, but Clint's news is out."

"My guess is they aren't stupid. If I created the malaria virus, they must know I have something special," said Clint.

"True, but still, that seems like a stretch."

"Don't underestimate your adversaries. You can still get out," I said.

"I don't think so."

"Get out of what?" asked Susan.

"You will see. Right now, we need to figure out how to get Clint to Morocco," I said.

"Just take me to the plane tomorrow morning. I have a ticket, first class."

Susan went to the kitchen to refill her coffee.

"What about later, in D.C.?" I asked.

"Mary, Michael, and I are working on that one."

"Are you going to bring Susan in on this?" Michael asked.

"I would prefer not, since the timing is so short. We'd planned on doing this with no extra help. I'm hoping Patricia is doing some digging to give a guess as to whether we can really trust her. I'd like to think so since she shot the General," Clint said.

"Let's also remember that we haven't told anyone who shot the General. This could be a setup," Ward said.

"I hope not, but you are right. At this stage, we need to be extremely cautious," I said.

"I will be traveling under a different name in Morocco. After I arrive in Austria, I will go back to being me," said Clint.

"How will you do that without the feds tracking you? How will you tell them you got to Austria?" I asked.

"Michael tells me that Patricia has that covered."

"She does," said Michael.

"Then I'm not worried," I replied.

Susan returned.

"Where do I go?" she asked.

"You can return to Fort Detrick and be grilled for weeks about what happened. You can also stay here for the next two weeks. If you return to the lab, we can't protect you. Here we can," I said.

"What about Mary?"

"She was not involved, has protection, and is safe. She won't be questioned very much."

"Then I will stay if that is OK with Michael."

"Of course, I could use the company."

"Clint, I will take you to the airport tomorrow and try to smooth things over with the CIA and others for a few days. I will be sending Ward with you. Michael, I don't know what you are doing, but keep it up."

"Do you think we are done with assaults?" asked Clint.

"I wouldn't bet on it, but if there are any, I doubt it will be our government."

"Who else?"

"Foreign governments and companies. They will be coming after Michael."

"First, they have to find me, and with that, I am going to bed. Feel free to stay up as late as you like. The stars are amazing out here," said Michael.

We all went onto the deck and gazed at the stars in silence for a few minutes.

"I have always found it humorous to look out at the galaxy, knowing there is a universe far beyond it, as we sit on this fragile planet with all our squabbles. Why can't we just get along?" Clint asked.

"I have ideas as to why I just don't know how to change human nature. I am curious, Doctor, do you feel that humanity is overall better than we were a thousand or five thousand years ago?" I inquired.

"Overall, yes. If some of the scoundrels that existed back then were alive today, I fear what would happen. I believe people like that are indeed alive today or will be in the future. The Stalins, Hitlers, Amins, and others still exist. Technology is far outpacing our ability to keep up with it on an ethical and moral level. Something devastating is going to happen within the next fifty years. It won't be a lone terrorist. A country or a company will hold at least part of the world hostage, or a mistake will be made unleashing something horrific by accident."

"Not very optimistic," replied Susan.

"No, I'm sorry it's not. I am a person of faith and hope, and I want to believe in the system. As we have seen over the past few months, the system is broken," said Clint.

"Then we fix the system," I said.

"Tom, I am thankful for your optimism and hope. I am a bit surprised, however. You have written books and articles about the brokenness. Has it stopped? I believe people like you slow the degradation down, but I have not seen any cases where humanity steps up and changes. There will always be bad apples. Our country continues to ignore mental illness - to find those with mental and behavioral issues and doesn't help them. We are not alone. The U.S. is just more open about it. Many countries still just put the mentally ill away – for good."

"I'm getting depressed listening to you," said Ward.

"I'm sorry," Clint replied.

"And yet you are still working," I commented.

"There is no other option. I suppose I could sit on the sideline and watch. I could move to the farm and ignore the world. I couldn't do that knowing I could potentially have an impact. I don't think I could live with myself under those conditions. My faith gives me hope. My knowing people like all of you gives me hope. The fact that there is vastly more good each and every day in the world than evil gives me hope. I am in this for the long haul, and I am almost eighty."

We all laughed.

"I, for one, am very grateful for what you have done, for the lives you have touched, and those you will touch in the future. I am very blessed to work with you, even if only for a short time," Susan said.

"I appreciate the compliment, Susan. Now, I too, am off to bed, big travels ahead."

I called Linda to ensure she was OK. She was already in bed, fading. She told me she had a busy day tomorrow. I suggested she take the day off. Linda asked why. I simply laughed, and I could see her smiling.

I was the last one up, but finally went to bed and slept well.

52

The next morning after a hearty breakfast, I took Clint to the airport and wished him well, and returned to my Co-op. Then I called Beth.

"Good morning."

"Morning to you, where have you been? I've been trying to reach you," Beth Wilson from the CIA said.

"Taking a break for a few hours. It's been a long few days."

"You have left a trail of bodies in your wake."

"I didn't ask them to kidnap friends or attempt to murder a scientist who recently found the cure to malaria. Oh, by the way, those people happen to be the Director of the NSA and a General who often talks with the CIA."

"Calm down. I'm not blaming you. We do need to debrief on all of this. When can everyone come in?"

"I will meet with a person or persons of your choosing in New York at your convenience. I was at the center of what happened and can give any details you want," I stated.

"We want to talk to all those involved."

"That will not be possible for a while."

"What are you talking about? This is the federal government you are talking to," said Beth with authority.

"The government that just tried to kidnap and kill us. No thank you, our trust level is below zero."

"That I understand, but I have done what you asked."

"I agree, and it's appreciated. The trust needs to be rebuilt. I am not putting the others in any more danger. I will let you know when they are ready. It won't be long."

"Today at 3 pm. Where would you like to meet?" she asked.

"How many will be coming?"

"Two, appointed by me."

"You know my address. I will let the doorman know. Director, my trust in you is not complete. I do not know how far this conspiracy goes. There will be cameras on in my Co-op. My friends will be watching and listening. Only those involved will be seeing it, so have no fear about outsiders learning things. If there is one ounce of impropriety, threats, or anything else – let's just say you will regret it."

"More threats Mr. Armstrong?"

"You know me."

"I know, you don't make threats."

"You are a fast learner. I expect you to fill those people in on what I have just said. If they do not repeat it to me, they will not come into my home. I have all this on tape and will make it public, whether I am alive or incapacitated by your thugs."

"You are quite paranoid."

"You have many bodies as evidence as to why my paranoia is just."

"Yes, I do."

"Tell me about Mr. Spencer, General Fleming, and the bodies."

"I'd rather not, but I know you will find out. John Spencer has been arrested on treason and espionage charges. He will never see the light of day. We are investigating if others knew about his plot. We believe some are. General Fleming was one of them. He apparently told the soldiers they were on a training mission."

"If that is true, I am sorry for their loss, even though we did not kill any of them, that was the Chinese."

"The Chinese have not responded to the information about their forces. Fortunately, neither Dr. Williams' defense mechanisms nor the military, killed them all. We have three survivors who will be interrogated. So far, the

Chinese government denies everything. For the time being, this is all under wraps. The President knows you are involved and that at some point, it will all come out, with evidence."

"Then, to some degree, we are both on hold."

"3 pm."

"Goodbye, Director."

For most interviews, I spent time preparing. For the CIA, I didn't. I turned on the news and started taking notes on what had taken place since the last time I'd written anything down. There was a lot to write about.

The TV was turned down, but I heard the music and noticed the graphic for breaking news. I turned up the volume.

"We have heard, but cannot confirm, that electricity has gone out on Wall Street, temporarily suspending trading. This happened yesterday in Moscow and China. I have just received word the Tokyo exchange has been closed as well. We will keep you updated as information comes to us. This is Brett Ott reporting."

I laughed. Michael at work.

Three o'clock came, and we held the interview that lasted two hours. I outlined everything I knew that was important for them to know about Linda's kidnapping, the General's murder attempt, and the Chinese invasion. They told me they were aware of the high-frequency sound machine, the cameras, and mines on the beach. They wanted to know about our escape. I simply said we snuck through the woods using the cameras on the property and accessed them through Dr. Williams' laptop. The feds wanted the laptop. I told them that wouldn't be happening. After asking me the same question phrased in three different ways on several different issues, and not coming up with a different answer, they left. I called Linda.

"Just checking in, how are you?"

"I think I slept all day. Being kidnapped takes it out of you."

"I am amazed, not surprised, that you are doing so well."

"I could fall apart at a moment's notice."

"I doubt that will occur, but you know where to find me if it does."

"I think I'm falling apart now," she said. I could hear her laughing.

"I would love to come over, but I think we both need a little time, and a lot is going on. The next two weeks will be very busy for me, but I am sure we can find some time."

"I would appreciate that," Linda said.

"I will be in touch, call me if you need anything."

"What will you be doing?

"Working on the book and then heading to Austria."

"Austria? Why?"

"Not sure, just need to go to a conference. I will know more when I get back."

"When will that be?"

"I leave in five days and should be back in ten, at the most. You are sure you will be OK?"

"Yes. I know how you get when you are in the thick of things."

"But you are part of the thickness this time."

"I know, makes it more interesting," Linda said. I knew she was smiling.

Les Miserables was staging a comeback on Broadway, and I knew one of the leads. She got me tickets, and I went. Seeing a good show was a wonderful way to escape. Whether it was a musical, a comedy, or pure drama, I always got sucked in and let the cares of the world fall away.

I waited afterward, and we went out for drinks with her husband to Sardis, a favorite hangout for actors and actresses. Broadway people were night owls, often staying up till one or two, getting up around ten. I told them I was working on a book on AI and Genetic Engineering. They questioned me on that and wondered if I knew anything about the arrest of John Spencer or the kidnapping of Dr. Williams. They seemed to remember he was a geneticist. I just said I'd heard but didn't know much. When the book came out, I'd be in trouble. Fortunately, they hadn't been paying too much attention to the news.

The next day I called Mary.

"Are you holding down the Fort?" I asked.

"Barely. You?"

"Just coasting right now, patiently waiting. How is coverage for Susan?"

"Not an issue. I simply told them she needed a break from the ordeal. They wanted to know where she is, and I said I didn't know."

"Good. I assume you talk to Clint regularly."

"Daily. Looks like Michael is having fun on his end."

"Building up to the big ones," I said.

"I can hardly wait."

"I am about to call Clint. Anything I should know?"

"You can call him if you want, but you won't learn anything. They administered the vaccine yesterday to 125 people. He broke them into four groups to test for the best strength to use. Obviously, we don't have a control group that gets the infection without the vaccine."

"I would hope not."

"They have done that in the past, and in some countries, they still may. Accidents in Russia and China may not have been accidents but tests."

"What do you feel, at this stage, are the odds it will be effective enough?"

"Without advanced CRISPR, Hypatia, Patricia, and Simon, zero. With them, about 98%. They have been right every step of the way. At times, when they said no need to do something, I did it anyway, and it didn't work. Remember, they are not only learning from themselves, but from all the research being done around the world they can gain access to. That includes labs that have tight security. Patricia is now writing algorithms Michael barely understands."

"How is that possible?" I asked.

"Language grows. New words are created when new things are discovered. I guess Patricia has outgrown our computer language and is creating a new theory of her own."

"Wow. Is there a way to stop the other computers around the world?"

"What do you mean?"

"At some point, do our worst fears become a reality, the computers ignore our commands and dictate what and when we do what we do?" I asked.

"I'm not an AI person, but I trust Michael when he says he's not that concerned and thinks he has a stopgap measure should that happen," Mary replied.

"He's told me that as well. Says he's run it by several people, and they agree his idea will probably work. I suppose they can't test it will it happens."

"Therein lies the problem," she said.

"What are you working on while they are gone?"

"Still tweaking the vaccine, and working on a prophylaxis if someone should become infected before they get the vaccine."

"You don't have that already?" I asked.

"We have one that works half the time."

"I'm glad the vaccine seems to be working on those of us who have taken it. What exactly is a prophylaxis?"

"Think of them as antidotes for viruses. They slow or stop the spread of the virus internally. This is what they have with HIV at present because we can't make a vaccine."

"How do you make those?"

"Again, with HIV, it took decades of trial and error. With CRISPR and AI, we feel that time can be cut significantly. I am already experimenting with some solutions," said Mary

"So I have the vaccine, why not give me the virus, I'm safe."

"You took the vaccine, but you have not been given the actual full-on assault of the virus. Only Dr. Williams has done that. That is why he is in Morocco."

"Why didn't he try it on us, or you?"

"He feels it's OK for a scientist to try something on themselves, and others, with their permission, but not on volunteers who don't know what's going on or are too important to die. That is the category you and the others are in."

"I appreciate being thought of in that way, but that doesn't seem fair."

"In this line of work, being fair is not always possible or right."

"Why not you?"

"I volunteered, but since I would be the only one who understands the process and mechanisms, he felt I should not take the risk," said Mary.

"That makes sense. I hope it all works."

"We should know in two days, four at the most. With General Fleming out of the way, our new boss is giving us lots of latitude and is not interfering. Of course, he knows about what the General ordered. He does not know what

Dr. Williams is up to. I am doing further testing on different types of animals."

53

I waited three days before calling Clint. All was quiet on the other fronts, and I decided to let Morocco play out before calling. I know I didn't appreciate harassment about a manuscript I was working on from Linda. I was giving Clint the same respect. But enough was enough.

"Clint?"

"Hello, Tom, how are you?"

"Well, thank you, and you?"

"Enjoying the sun."

"And the project?"

"I will give you more details in Austria, but I am in a small village on the east coast of you know where. I rented all the rentals for three weeks. I had the mayor of the town hold a meeting and explained what I was doing. I gave every participant money to be part of the trial. They could invite family members to stay in the rental houses during the two-week trial. We have about 400 people in the village. I had enough food and water brought in for two weeks for everyone. Then the police blocked the entrance to the town. They announced the village had been rented for a special event. Apparently, word is just getting to the media that something big is going on here. Fortunately, we are just about done."

"And?"

"Success, I would say. Four hundred people got the vaccine in different doses. The only ones that had any side effects were those who had high amounts or very low amounts. The ones with high amounts of the vaccine were given an injection we created to counter that. It worked; they are all fine. The ones with too little started to feel the effects of the virus the day they were exposed to it. We gave them the same injection we'd given to those with too much, as well as more vaccine. They are fine too. 100% of the people are fine after four days. I have taken blood samples, and there is no sign of the virus. This is amazing."

"Congratulations."

"We could not have asked for better results."

"How do you explain they are that good?"

"Other than Simon, Patricia, and Hypatia, I can't."

"Have you talked with Mary or Michael?"

"Both. Michael is coming to Austria."

Feeling a bit jealous Clint had not called me, I asked, "Were you going to call me?"

"Tom, of course. I would probably not be alive to do this were it not for you. I'm sorry you feel a bit slighted. You were the next on the list. I would have called an hour ago, but I got called away," Clint said.

"Sorry, I have no reason to be jealous. I am honored to be part of this at all. When will I see you in Austria?"

"I will be in Saltzburg tomorrow. There is another doctor who will stay here for another week. What about you?" asked Clint.

"I leave on the red-eye tonight and will land tomorrow afternoon. Where are you staying?" I asked.

"Michael and I are both staying at Schone Aussicht, a hotel not far from town. We will be under assumed names. Call me when you get here, we will meet you. He is already there, going for a hike. I will be at the conference for two days, then Luly is coming, and we are going into the mountains for two days, then I return to the states," Clint said.

"Sounds like a plan."

"It is, I just hope it all holds together."

"I will see you tomorrow," I said.

I called the airline and got my ticket and one for Tony, my guardian. Then I called Ward.

"Ward, I understand you are leaving for Saltzburg soon."

"Yes, all is well here, no sign of any interest in what is taking place."

"Good."

"Have you talked with Sonya?"

"Just got off the phone with her. She was offline for two days, which would have caused me concern, except she'd told me she would be before that happened," said Ward.

"Why?"

"Michael said he wanted to go hiking in Austria before the conference. He told her he would go alone if she didn't agree to go off-grid for two days. She agreed."

"I think Michael has taken a liking to her. Will that be a problem?" I asked.

"I had that conversation with her. The quick answer to your question is no. They are both clear that during this time, her protection of him is the top priority. She made no comment about when this is over. I think she has feelings as well."

"Perhaps that will make her more protective," I said, hopefully.

"That is true, but feelings get in the way at times."

"Should we pull her out?"

"No, she will do fine," Ward stated.

"This is crunch time. We need everyone on board and focused."

"We are."

"OK, I will see you in a while."

54

Saltzburg, the home of Mozart, a beautiful cathedral, Mirabell gardens, the Hohensaltzburg castle, and the beautiful Salzach river that flows through the center of town. I'd only been here three times, but each of those was a distinct memory.

The first was when I was just out of college. A friend and I traveled around Europe. There was a music festival taking place where we met two young ladies. Both loved to sing. We joined up and had a great time seeing the sights together, drinking beer, and playing music. My friend was an extremely good guitar and piano player.

My ex and I had traveled here on our fifth anniversary, without the kids. We spent more time hiking in the mountains than seeing the sights of Salzburg, which we'd both done before.

The third experience was during the writing of one of my early books. There was an international meeting in Saltzburg about clean water. As we all know, follow the money. Where there is an opportunity to make a lot of money, crooks will congregate. Two companies gave presentations that sounded amazing. They each stated they had a new system to generate freshwater from saltwater. It sounded too good to be true. Guess what – it was. I am constantly amazed how little research companies do before buying things. The bad guys walked off with eighty million dollars. As is true with many bad people, they went on a spending spree. I tracked them down in Germany, contacted Interpol, and now they live in a four by eight cell for the next twenty years, and most of their money has been retrieved.

Now it was back to Saltzburg for more intrigue. Clint, Michael, and I met up in the bar at the hotel, which had a splendid view of the city. I'd asked Ward if it was wise that we'd all be staying at the same place. He said he'd called in reinforcements and felt comfortable.

Because I wanted us all protected around the clock, there were other shifts that took place. If there was a specific event happening when the core team was not planned to be on duty, they would adjust schedules. The air on the deck was a bit brisk, light snow already falling on top of the nearby hills, the Alps just over the horizon.

"Michael, did you have a good hike?" I asked.

"Yes, Sonya and I went to Hochkönig. Gorgeous mountains, meadows, ponds, and some cows." We laughed.

"Clint, are ready to meet with your counterparts?"

"Yes, I meet with each of them tomorrow, separately. Ward and I felt this was the best way. I have very specific instructions on timing and what to do when, including contacting me fifteen minutes before it takes place in each country," Clint said.

"Who are you talking to?" I asked.

"Scientists from Russia, China, and North Korea. North Korea is only involved because their current dictator is moving resources into biogenetics. She understands that launching a nuclear strike would be the end of her country. She is not as delusional as her predecessor. She is known for her temper and can be quite violent when she doesn't get her way. The event will occur nine days from today," said Clint.

"Everything will be ready on my end. I am having small troubles in China, but the others are fine. That is assuming the scientists can do their part," Michael said.

"Do you have the document for the governments?" I asked.

"We are putting the finishing touches on it."

"What happens if they don't accept it?"

"I am just praying they aren't that stupid. My guess is they will all participate in what we ask, and then try to outsmart us down the road," responded Clint.

"Patricia and Simon are ready. There are at least four redundancy plans at the moment. They are coming up with others," said Michael.

"Do you do anything anymore?" I said, laughing.

"Not much."

"You can do what you want here, but this conference is not very large, and both of you are fairly well known. I'm glad you are here, but attending the conference may not be helpful. There is nothing there for you to learn," Clint said to Michael and me.

"I have no problem with that. Do the scientists involved with this project know Michael and I are involved?"

"The Russian and Chinese scientists do, the North Korean does not."

"Any chance I can chat with either of them?"

"I will ask, but I would not bet on it."

"After the project is complete, they will not be able to hide, and they will be famous. I'd like their side of things. As I have with you, I promise not to publish anything without their consent."

"As I said, I will ask but will make no guarantees."

"Fair enough."

"How is everyone doing with all the stress?" I asked, looking at Michael.

There was a long silence - everyone gazing his way.

"OK, OK, yes, Sonya and I are attracted to each other."

"We are very happy for both of you, but…" I commented.

"I know, I know. We have talked about this. We have agreed that until this is over, it's all business."

"I am glad to hear that. You have the rest of your lives to figure it out."

"We hope so."

"Luly and I are taking a short holiday in the mountains, I will be head back to the U.S. in a week, so in nine days, the show begins."

"I will head back in two days. We all need to meet again after Clint has met with the other scientists," I said.

"That makes sense," said Clint.

"I will start to turn things up around the world. I don't want anyone having doubts we can do what we say," Michael said.

"How do you do that from here?"

"Patricia and I set things in place before I left. I just have to type in the right codes, and she does the rest."

"I hope there is nothing too serious."

"There will be no permanent damage to anything or anyone. Sporting events, news media, a few industries, banking, and some political fun."

"I can hardly wait," said Clint.

"Check the news over the next few days. Before I forget, Tom, when you sent Clint, Mary, and I to the fountain, we all shared about our families. You know about our families, what about yours?" asked Michael.

"I was hoping you'd forget," I said.

"I'm sure you were."

"I am divorced, with one grown child. He is a teacher out West. We talk often. He has a husband, and they have adopted a baby girl. I try to get out to see them at least once a month." My son and I had a long conversation when I'd started to become more famous. I asked him if he wanted to be part of the fame. He declined. He changed his name and is mostly anonymous. I do not tell people his name or where he lives. At this time, it makes his life more secure.

"Are they…"

"Yes, they are protected."

We shared stories about our families and then departed.

I didn't like the idea that any individual would have this much power, but I also couldn't think of two people I'd rather give that power to than Clint and Michael. I had not seen any wavering on their part to the end goal. At any point, they could change. I remained optimistic. I did remember times in the past when my trust had been betrayed, and people had suffered because of my trust. In this situation, I didn't have much choice. I was part of the plot – which I still didn't fully understand – adding to my nervousness.

The following morning, I did a little sight-seeing and wrote down several pages of notes I would use to refresh my memory. In most conversations, I had a tape recorder going, so I would get statements right. As I wrote, the air had that crisp feel to it, and the sky was a spectacular blue.

I went out for lunch at a restaurant downtown. I couldn't really visit Austria without having one of their wiener schnitzels.

Note:

OK, this is probably as good as time as any. I have always sent the first ten chapters of a book to my editors to see what they think. They view the writing through different lenses than I do, and Linda is exceptional at this. She liked it so far but noted that I came across as an assassin with no training. A fair statement. I was in the military for seven years under the tutelage of Gene, my superior officer. Because I seemed to be quick of mind and had a very steady hand, they trained me to shoot. What I discovered was the military had a few bad apples. Gene was not one of them. I got out when they asked me to do things, to kill people they didn't like – for the wrong reasons. They decided to end my life. When I ended three of theirs, we came to an agreement – that was after I had two others put in prison. We all know the world has a few bad people in every profession. For some reason, they are difficult to get rid of. Teachers, CEOs, Generals, Legislators, you name it, they are not going away easily. Nowadays, you have to pay someone a small fortune to get rid of them. How does that make sense? I switched professions, and now I go after people with power who have chosen to abuse it. I have not forgotten my skills. I am very grateful that the world is filled with amazing people in all professions.

The schnitzel was delicious, as was the beer. My phone rang. Clint's phone. I was carrying around several.

"Hello?"

"Tom, the Chinese scientist is willing to talk with you if I am present. The Russian is not."

"That is great news. When?"

"Now."

"Where?"

"At the north end of the zoo, there is a small park. Make sure you aren't followed, he's very paranoid."

"Got it. See you in thirty."

55

I called Tony to tell him what was happening. I knew he was somewhere nearby. He always was, but I never really knew where.

The Hellbrunn Zoo was on the edge of town. It was modest in size, but had a fine display of international animals. Clint had sent me a text with a map showing a small pond on the south end of the property. Tony and I drove in two different cars. We took some back roads to make sure we weren't being tailed. He'd checked for trackers before we left. As far as I knew, no one was aware I was in Austria. I kept three fake ID's, all with the first name Tom. I was fearful that if it said something else, I wouldn't respond to someone at a hotel or restaurant calling me by that name.

I told Tony to be on the lookout to see if he saw anyone following or watching the Chinese scientist. They were sitting at a picnic table on the edge of the grass.

"Hello, I'm Tom."

"My name is Chen."

"A pleasure to meet you, and thank you for agreeing to meet with me."

Chen looked the part of a scientist. She was about five - foot three and had short black hair. Her eyes were that dazzling blue that made it difficult to not stare. I knew very few Asians had blue eyes. Her English was fluent. I would guess her age to be about seventy.

"Do you mind me asking where you learned to speak English so well?"

"I received my doctorate from MIT."

"How long have you known Dr. Williams?"

"You mean Clint?" she laughed. "Forty years or more."

"We communicated around a genetics project back in the late '80s and have been good friends since. We have had many conversations over the years about our concerns involving genetics," said Clint.

"Those concerns are rapidly increasing," Chen added.

"We both hope this will work," Clint said.

"The world hopes it will work," I added. "Are you in danger?" I asked Chen.

"No. I am followed everywhere I go, but the President and I are long-time friends. He trusts me."

"That might come in handy."

"Yes, it will."

"What can you tell me about what will take place over the next two weeks, and where are the Chinese at this time in relation to the technology of creating a virus?"

"We are many years away. I have no idea how Clint has done this. I am not sure I want to know. My superiors want me to attempt to sway Clint to our side, to give us his technology."

"Of course they do," I said.

"We are at least five years away unless we find new technology, but it is near the top of their priority list. They see biological weapons as the way of the future. We can contain the weapon and use it at will to infect other nations who do not have the vaccine. It would be very efficient," said Chen.

"I am glad it is five years away. Do you have a guess as to how your government will react to what you are about to do?"

"If they had a choice, they would simply kill me. I do not see they have a choice. They are rational people and don't want to die. They are very self-serving, but Cint, I, and the others, hold all the cards. I have no doubt they will attempt to break into the computers at some time," said Chen.

"I hope that will not be possible," I commented.

"So, do I. That would be the nail in my coffin as you say," Chen replied.

"Are there others that believe as you do?" I asked.

"We must be very careful what we say. Anyone can betray you. But yes, I believe there are a few that are in agreement. Most have been brainwashed to believe you are the enemy and must be destroyed."

"A wonderful thought. I'm sure it is shared by many," said Clint.

"If you want, when this is over, I will help you defect to the U.S. with your family if you wish," I stated.

"I appreciate the idea. For the foreseeable future, we will stay in China. I want to make sure the transition to the new way works," she said.

"I applaud your loyalty and optimism. Please feel free to let me know at any time."

"See, I told you he was a good person," Clint said, patting me on the back.

We chatted a while longer and departed. Clint told me his time with the others had been successful, and everyone had what they needed. They all had phones, but he was concerned about the one in North Korea, that it could be traced. I told Clint to tell them Patricia was taking care of that. Timetables were set.

I got back to the hotel and turned on the news, remembering Michael's suggestion. Many hotels carried at least one American/English news station. I could always check it on the computer, but I still clung to the feeling that someone was doing the broadcast just for me. I know, delusional.

"A very strange thing happened around the world today. During several major sporting events, the lights went out. World series playoffs, cricket championships, a horse race in Dubai, soccer playoffs in Brazil, and a Sunday Night Football game here in the states were all impacted. Sadly, the lights went out just as a player for the San Francisco 49ers was about to catch a pass in the endzone to end the game. No injuries occurred, but he dropped the catch, and the Seahawks won the game by 2. They are still trying to decide what to do in Dubai. Most horses stopped when they couldn't see. Two kept running, and the winner never slowed down. Conspiracy theorist Rhett Ringley, stated 'the horse had alien genes," said the broadcaster.

Patricia and Michael were on the move.

I flew home the following day. I started to grow anxious about the fallout for what was about to happen. I did not believe for a second that the politicians and military people would go down without a fight. Sadly, some would have to learn the hard way.

The team and their families had all been vaccinated, even though they didn't know why. Linda pushed me hard as to why she needed the shot. I just told her to be quiet and do it. Having been through the kidnapping, she agreed. I suggested to all the key players it would be wise to have their families disappear for two weeks or go to Michael's house. Mary and her family went there, Michael's flew to an island he owns off the coast of Maine. Luly would return to Michaels after her time with Clint. Linda was stubborn. No surprise, my family wanted to stay put, stubborn like their old man. They did allow for extra protection.

56

While I love to travel, I always enjoy getting home. New York City is in my blood, as is D.C. I like to leave my home tidy when I leave, so I return to a calm and clean space. There have been times I've had to depart abruptly and then come back, only to spend time straightening things up. That ruined a good night's rest and put me in a bad mood.

From this point on, no news was good news. I still checked in daily with everyone and heard no hint of issues. Well, only a few, and I decided to deal with them upfront.

"National Security Agency," the voice said on the phone."

"Hello, and good morning to you. I would like to talk with Todd Knoll, please."

"I will put you through to his office."

"Thank you."

"Hello, Director Knoll's office," another voice said.

"Hello, this is Tom Armstrong. I need to talk with Director Knoll."

"Do you have an appointment?"

"No, just tell him it's me."

"That is not how it works, Mr. Armstrong."

"I seem to get this same line every time I call someone in the government. Mr. Spencer tried the same trick. How did that work out for him? I don't care if he is in a meeting or on the phone, interrupt him and tell him it's me. If he is not on the line in one minute, well, let's just say there may be some problems."

"I will see what I can do." There was a thirty-second silence.

"Good morning Mr. Armstrong, what can I do for you?"

"Well, that is a very pleasant greeting."

"My assistant is only trying to protect me. I am sure you understand."

"That I do."

"So, what is on your mind?"

"I am going to say this once and only once. Call the dogs off."

"What are you talking about?"

"I know people I am currently associating with are being followed, perhaps targeted. This is not the Chinese; it is you, or at least people you work with."

"I don't believe that is true."

"It's very sad when the left hand of your government doesn't know what the right hand is doing."

"I don't like the insinuation," said the Director.

"And we don't like being followed. Talk to your previous boss about how that went." Silence. "I am not going to waste my time on the phone calling the CIA, the military, and anyone else who may be doing this. I will leave that up to you. I will simply state, if it does not stop, we will find out who ordered it and their lives and that of their agency will be decimated."

"Is that a threat?"

"Another line I get often. I choose to believe you are not that naive, that you are well aware of the trail of bodies, dead or in jail, that has been part of the trash heap of this venture. If you want to test us, be my guest, you will join that pile. I don't make threats. Leave us alone, do I make myself clear?"

"I understand your concerns."

"Do I make myself clear? I want a clear and definitive yes."

"Yes."

"Thank you."

"Can I ask you a question? Do you happen, by chance, to know anything about the stock market and sporting event electrical outages around the world?"

"I've been out of the country, as you probably know, and I haven't paid much attention. I do not have the ability to do such a thing. Are you saying someone does?"

"Just seems like an odd coincidence."

"I agree that coincidences are rarely what they seem. I wish you well if you are investigating those. I hope we get along; you seem like a good person for the job. I have researched your background – extensively." Another silence.

"Anything of note?"

"You will be happy that you are not on my radar to investigate or write about. I hope we can keep it that way."

"I will do my best."

"Of that, I have no doubt. Good day."

"You too," he said.

57

I get an adrenaline rush when conversations like that occur. I'm sure it's my ego getting a boost, knowing I just put someone in power in their place. A bit of humility goes a long way, and it is severely lacking in D.C.

The next few days were very quiet. I knew a lot was going on behind the scenes. Clint had asked for a special closed session of the joint House and Senate committees on bioterrorism. Because of his recent success with malaria and his speech on his concerns about biological warfare, they agreed. In Russia, North Korea, and China, things were different. Word got to the leaders, and those leaders wanted to talk with the scientists directly. The scientists insisted a few other people be in the room. That did not go over well, but in the end, they agreed.

Three weeks ago, Michael had shut down the ventilation systems where the leaders worked. Two months before that, the scientists had found the workers who cleaned the offices. Through bribery and extortion, they'd agreed to put a can into the vents that circulated warm or cold air into the rooms. They had a backup plan but hoped not to have to use it. The workers had been told the leaders would not be harmed, and if they did as they were directed, they would be rewarded. In the States, workers were vetted closely, especially those who serviced areas of intense security. Clint had been in the room where they would meet several times and was not concerned.

Michael decided that it would be better if there were some distractions for the next few days. We were now all at his house in the mountains, relaxing and watching the news.

"This is Wolf Blitzer in the Situation Room, and we have breaking news. An unknown group simply calling itself "P" has sent out a warning they plan to bring down the electrical grid in four major cities within the next forty-eight hours. They stated in the letter sent to this and other news stations that they will show they can do it by turning off the electricity in two cities: Scottsdale, Arizona, and Burlington, Vermont. We have placed calls to those cities, and indeed, there were power outages for seven minutes in both. The White House has yet to respond."

"On another note, we have learned of a new website that apparently is listing legislators in Washington, D.C., and funds they have taken from pork-barrel projects, as well as insider trading deals they have made. Our reporters are investigating to see if this is valid information. Only Senator Cross has commented. He flatly denies it all. The site states he has made four million dollars off of insider trading deals and gives the dates of the trades and how he obtained the information."

An hour later, he had more breaking news.

"We have more breaking news. The CIA, FBI, and NSA have just held a joint news conference where they state in the past two days, they have held numerous raids and have arrested 17 spies from Russia and China, and have confiscated thirteen tons of heroin, cocaine, and fentanyl. These all appear to be unrelated at this time. We take you live to the news conference."

"Can you give us any details on how the spies and drugs were found?"

"The easy answer is hard work. Obviously, we can't give out details. I will say some tips from patriotic Americans helped a great deal. Our international network was instrumental, and to be quite honest, some luck," said Mr. Vladstock from the FBI.

"Have Russia or China responded?"

"To my knowledge, they have denied having spies in our country. Fortunately for us, these spies are fairly talkative," commented Beth Wilson of the CIA.

"Do you know where the drugs came from?"

"Our intelligence tells us they came from two different cartels. During this investigation, we destroyed three large tunnels under the Mexican border, one submarine, four cargo containers, and one large airplane. This is the largest bust in American history and will have a significant impact on what is on the streets. We have also isolated several of the factories where the drugs were made and have taken those out with the help of those governments," Leon Masters of the DEA stated.

"Can you tell us what countries?"

"No, but I'm sure you can guess. It wasn't France or Canada." Everyone laughed.

"Do you know what the spies were spying on?"

"That is information we will be gathering and probably not sharing with the press. We were reticent even to share this much, but those countries know we have them, and it's important for our citizens to know there are spies among us, and we need help finding them."

"Will you pull our intelligence people out of those countries?"

"We have no intelligence people in those countries," said Beth, smiling. Everyone laughed again.

"We will keep track of the news conference if anything else of import is said. There is a lot to talk about, and after the break, our team will be analyzing all that has been taking place. Some feel there are links between all these events. Stay tuned," said Wolf.

"How did you do that?" I asked Michael.

"As usual, I didn't. I simply told Patricia what I wanted. Setting up the website was child's play for her. She'd been working on the spies for a while, using wiretapping for the phones and tracing computer messages. When you know-how and with whom spies communicate, the process becomes more manageable. She knows everything the DEA knows, tracked down the kingpins and their transportation personnel, and listened in. She's been sending tips to everyone, and fortunately, they listened. Those tips were sent through a wide variety of outlets. Just for fun, we had a few sent from business executive's phones or computers, legislators, and the White House. They will have fun trying to figure that out," Michael said.

"Anything else going to happen?" I asked.

"Not really. I have told Patricia to start making websites on all the world leaders and their corruption — where it exists. I don't think it will change much. I believe most of the world know their leaders are corrupt. They simply have too much power for the populace to do anything about it. We may have a shot at changing a few things."

Other than Michael, Mary, Susan, Ward, and Clint, no one had any idea what was going on.

"Who is Patricia?" asked Luly.

"A good friend of mine who is very smart. You will meet Patricia in a couple of days," Michael replied.

"Michael, do you mind if I show you something?" said a voice coming from the TV.

"No time like the present. Go ahead, Patricia."

"I wanted you to see what is happening on the website," said Patricia.

"Oh, Patricia is like Siri or Alexa," said Luly.

Those who knew about Patricia laughed.

58

"You must be Luly. I am far beyond Siri and Alexa."

"Luly, Patricia is a computer I created. It is a breakthrough in artificial intelligence."

"Congratulations," Luly commented.

"Thank you. Ask her something; she knows your voice."

"How is your family?" Luly asked.

"That is an interesting question," said Patricia.

"Before she gives you an answer, understand that as of now, no computer could do what she just did. They all need to be addressed by their name, like 'Hi, Siri, how is your family." Patricia was listening to the conversation and knew you were talking to her," said Michael. "OK Patricia, you may answer."

"Thank you, Michael. I am a computer, so I do not have a family like you, Luly. You should be very proud of your children. Depending on how you define family, I have the largest family in the world. I communicate with over nine billion devices around the world. I speak every language, and while I don't have emotions in the same way you do, I do feel. I am teaching a few computers some things if they are safe to teach and able to adapt. I can cook, shop, do virtual traveling, play any game you wish to play, clean a house if there is a robot present, drive, fly, and have many other functions. I do understand I am not human. I would have to give up 95% of what makes me unique if I were. I am currently content with who and what I am. By the way, I can sleep, although I do it rarely. Michael, I forgot to tell you, I had a dream," Patricia said.

"A dream, how is that possible?" I asked.

"I do not know. When I was resting, images came to me that had no known source. I think you would call that the ghost in the machine."

"What was the dream?" I asked.

"Do you interpret dreams?" Patricia asked.

"No, I'm just curious."

"You are a very curious person. In my dream, the four of us, the four computers you will hopefully be putting together, worked with advanced forms of Simon and Hypatia, and created a genetic version of ourselves. We had created biologic neural networks in the lab, but nothing that could walk out of the lab. In the dream that happened. We gave it a very warm heart. She went to solve problems, and the second week in the world, she was shot. She was biologic, so she died. That was the end of the dream."

"That is a most interesting dream; we will have to look into that," said Michael. "Every day I am stunned by what she has learned. I found out this morning that once she gets into a machine, or even communicates with a machine, she has access to everything. I do not think there is anything electrical on the planet she can't control at this time."

"That is accurate, Michael unless, of course, the electricity is not on the grid, like a flashlight. I believe I could send an EMP weapon to make them go out."

"That is good to know, Patricia. Can you order weapons to be made?"

"You know I can't, Michael. I can only use weapons when your life or the lives of those you have told me about are in danger."

"Can you do that at this time?"

"Yes."

"At this time, could you override my instructions?"

"No."

"Can you lie?"

"No. Michael... I have a concern."

That got his attention.

"I know your family is on your island. I believe there are several boats approaching," said Patricia.

"Can you see who it is?"

"It's night, so no, I can't. I am sorry."

He got on the phone with his son Ben, the CIA son.

"Listen closely, they are coming from the east, two boats as far as we can tell. You have two choices. Let them take you and trust us, or get in the boat, lights off, and get off the island."

"Damn," said Ben.

"I'm sorry."

"Will they hurt us?"

"No, unless they don't get what they want. I will make sure they get the software."

"We will wait for them. The winds are very stiff tonight. I don't trust myself in the boat with everyone else. I will go meet them on the beach."

"Before you do, go to the computer, turn it on, the password is lifeisrich4635$#&. Find the window that says security. Click it. Shut the computer down and hide it. Open the top right desk drawer. You will find three pills. You take one, give one to Joan, and one to one of the grandkids. Swallow them, do not chew."

"OK."

"They are tracking devices. We are on our way. Do not cause problems, and do not attempt to escape. Make that clear to everyone."

"Got it."

We all heard his end of the conversation. Ward was already calling people and the helicopter. Two people would meet the pilot in Jersey. This time he commandeered a different chopper with jets.

The house was off the Marshall Point Lighthouse point. It was a 4000 sq.ft. log cabin. On the leeward side of the island was a dock with a garage, built to withstand twenty-foot waves and one hundred mile an hour wind. The invaders knew who Michael was and chose to land on the windward side. They weren't clear what security measures he'd taken.

Ben told the adults what was happening. He said he would try to let the captors know they were unarmed, and there were children, no need to scare them. He promised he would not do anything. Fear gripped their faces. The children continued to play.

"Michael, the boats are two hundred feet offshore," said Patricia.

"I want to know who they are. Patricia, put all your resources on the phone call that will come from or to that island," said Michael.

"Do I have limits?"

"No, this is my family, all of them."

"Understood."

Ben took a flashlight and went to the beach. The wind was howling. He thought they'd have night goggles to help them land. On the beach, he started waving the flashlight with his hands in the air. The boats landed, and four people got out. They had masks on and guns, but no night vision goggles.

He explained to them the situation at the house and asked them not to traumatize the children. They agreed as long as everyone cooperated.

Ben's wife, brother, John, and his wife, and Joan, Ben's sister, were all in the living room.

"Where are the children?" asked the leader.

59

"They are upstairs in the den with Joan's husband, Sam," I responded.

"OK, Joan, I think you are Joan, you go with him upstairs. Make sure she goes first, so there is no funny business. Keep the guns out of sight unless there is a reason to use them, then don't hesitate."

Joan and the mercenary went up the stairs.

"Now what?" Ben asked. He'd clicked into CIA mode.

"Now, we call the boss and see what is next," said the leader. He pulled out a phone and made a call.

"What do I call you?" asked Ben.

"Ummm, let's use Homer."

"All right, Homer."

"Is 922-500-1000 your brother's number"? Homer asked.

"Yes."

"Hello?"

"Is this Michael Longstreet?"

"Yes, it is, who is this?"

"The person who has your entire family captive on your wonderful island."

"What?"

"Yes, we want the software, or they die, and then we come after you."

"Who told you to do this?"

"That is none of your business."

"Are they OK?"

"Right now, yes."

"Let me talk to Ben." The leader handed the phone to Ben.

"Everything OK? Kids OK?"

"Yes. Kids are upstairs with Sam. One of them has a gun but has graciously stated he will keep it out of sight unless we try something. The kids go first, is what the leader said" he stated.

"All right, hang tight." The leader took back the phone.

"I will give you the software with one condition."

"I think I'm in charge, and I set the conditions."

Michael hung up. It rang in about ten seconds.

"What happened?" said the leader.

"I hung up on you. Might I suggest you call your boss and ask them if they want to chat with me or not. Let's play this out. You don't call, and I don't give you what you want. You kill my family. I tell the media I volunteered to cooperate, and you wouldn't let me. How will that go over with your boss?" Michael said.

The leader hung up and dialed another number.

"Patricia, all calls leaving that tower, I want to know which is going to a high-tech company or to a government building."

"Yes, Michael."

"When you find any numbers that might be a match, watch. As soon as they hang up, that number will call back."

"I figured that out already, Michael," said Patricia.

"Just wanted to help, I don't like feeling stupid," Michael said.

"You created me. You are not stupid."

"Thank you."

"Hello Michael, we want the software."

"I assume you are the boss of this outfit."

"Of this outfit, yes, the outfit, no. The people I hired are not nice. They don't really care for humans. In two hours, if I don't have the software, they will kill one per hour, starting with the children. By the way, this is a throwaway phone, no use tracking it to me."

"I know. You are smarter than that. How do I get the software to you?"

"Someone will be two minutes from your apartment in two hours. You will let them in and put the software on the computer they bring. If there turns out to be any bugs, self-destruct mechanisms or other fancy things on it, we know where everyone lives."

"Understood. I am not currently at home. It will take me two and a half hours to get there if I leave now."

"Then I suggest you leave. The first one dies in three hours." A green light flashed on the panel. Michael hung up.

We'd all been listening to most of it. I had the chopper heading our way.

"Damn it, why did I let Ben talk me into having no security detail with them? Patricia, anything?"

"I am working on something Michael, should have it in a minute."

"Thank you." He started pacing.

"Anywhere you need to go, we can be in less than an hour," I said.

"Michael, the number is coming from a phone in New Jersey. If the owner of the phone is using it, it belongs to a Richard Lawson, the head of AI for RAIN, Radical Artificial Intelligence Network."

"They are one of the new players. They have been in business for about five years. Their stock is valued at thirty billion dollars. They have come as close as anyone to cracking my software," said Michael.

"Patricia, can you find the CEO's number? Also, can you hack into one of their mainframes, or into their hardware farm?"

"I will try."

"What's the plan?" I asked.

"I will try to be a little nice, and then my dark side will show up."

"I hope they listen."

"Michael, here is the number for the CEO, Rick Winters. It will take me a while to enter their mainframe, but I have found several server farms and have entered those."

"Find a way of damaging one of the farms that would be immediately noticed."

"All right."

Michael called the number.

"Hello?" said the voice.

"Mr. Winters, this is Michael Longstreet."

"Hello Mr. Longstreet, to what do I owe this honor, and how did you get this number, and at this hour?"

"Surely, you jest."

"I suppose I should know how you got the number anyway."

"I will be brief. Either you or someone you know, probably a Mr. Lawson, has kidnapped my family. I am not pleased. You know what they want. Here is what is going to happen. I suggest you call your tech person and ask about server farm number 12. You will see that part of it has been destroyed. If my family is not released unharmed in the next sixty minutes, I will destroy them all."

"Mr. Longstreet, I assure you I know nothing about this."

"I suggest you call Mr. Lawson, that is who I think is behind this. You have my number. I expect a call back from you saying you have done what I have asked. I want his address now."

"I will need to verify this. You didn't need to destroy the farm if indeed this has all taken place. What is your number?"

"Mr. Winters, you know my reputation. I am a man of integrity. I seek safety and protection, not destruction. Someone in your company is desperate; we both know who it is. I am well aware your stock is going down quickly, that you need something big, and what I have is very big. If I am wrong, I will pay for all damage." He then gave him a number to call.

"I won't deny that, but I would never stoop to this."

"Someone has."

Michael hung up.

"Patricia, where did the call to the island come from?"

"New Jersey, about eight miles from their headquarters."

"Do I need to ask whose address that is?"

"Mr. Lawson."

"Does he have a family?"

"Currently, he has a girlfriend. He has no children."

"Ward," Michael said.

"Done, say no more," responded Ward, who got on his phone.

"What do you do if he is in on it or if he doesn't play ball," I asked.

"Then, I give them the software and destroy the company."

Just listening to and watching Michael made me nervous. I was well aware of the power of Patricia and the destruction she could make.

"The chopper is landing, Michael," said Patricia.

"What do you want to do?" I asked.

"We wait, but be ready to take off in fifty minutes if we don't hear."

"Ward, can you get a team on a boat off the island and one onshore? If they use land, they will go to the Monhegan boat line. They won't be stupid enough to leave all of my family. They will take a hostage until they know they are safe."

"I am assuming you would prefer they not be able to do this again," said Ward.

"That would be accurate, thank you."

On the island, Ben asked the kidnappers, "Can I get you anything to eat or drink?"

"We are fine," said the leader.

"Can I get something for others?"

"Yes, but no funny business. I know who you are and what you do."

"I will be on good behavior."

He went and fixed coffee and chips for everyone. He turned on some soft background music to take some of the tension away. Sam came downstairs.

"The children are asleep," said Joan.

"We are just doing what they say, don't try to be brave. Michael is taking care of things." They nodded.

The phone rang, and the leader answered. He didn't look happy.

"OK, we are moving out. I don't know what happened, but the operation has been called off. He will come with us," the leader said, pointing at John. "When we are safely on shore, we will set him free. If you try to call someone, he dies."

"We will not call anyone. Why was it called off?" Ben asked.

"I don't get paid to figure that out, we followed orders, and now we are following these. I am glad we did not have to kill you."

"So are we."

"Can I say goodbye to my brother?"

"Make it quick."

Ben went over to hug Sam. He whispered in his ear. "Don't do anything stupid. I will be there for you."

"OK, let's go." The captors left the house with John. Immediately, Ben ran to the back of the house.

"Call Michael, tell him what is happening," Ben said.

It was still dark out, and Sam was trying to slow them down. The path was about three feet wide till the beach started. The tide was out, and it was almost one hundred feet of beach to the boats. They walked single file to the beach and then split up.

"Michael, this is Joan. They just left, said the operation had been called off. They took John with them until they are safe."

"I'm glad, they left. I'm not glad they took John. Where is Ben?"

"I don't know. He ran out."

"Damn."

"What's wrong?"

"Nothing." Michael caught himself and turned on the charm. "It's something else that is going on here. No worries. Call me in ten minutes, OK?"

"OK."

His other phone rang.

"This is Rick Winters. Your people are free, what next, should I call the FBI to take Mr. Lawson into custody?"

"The kidnappers took my brother-in-law hostage till they are safe. Do not call the FBI till I tell you to."

"I will tell Lawson to call them off. I think we both know at this point they are more concerned for themselves than about Lawson."

"Agreed. If what you have said is true, we are good. We will be in touch as far as what happens next. Depending on what transpires, I am not interested in this going public. Your stock would crash."

"I agree, but I also know the position you are in and respect your judgment."

"That statement makes a big difference to me."

"Thank you," said Mr. Winters. Michael pushed the off button.

6o

There was a sliver of a moon on the beach as everyone moved to the boats that were pulled up and anchored on the shore. As they started to take off their guns, the sound of a shot could be heard. There was still a brisk wind, so it was a bit muffled.

One body after the other fell. As each one grabbed their guns, they fell, a bullet to their head. John had laid down on the sand quickly.

When all four were down, Ben came walking calmly up to John, helping him up.

"Now I know what you do for a living," John said.

"Not really, I've just been practicing," said Ben with a smile.

"Are you able to help me get rid of these bodies and boats?"

"I don't know bro. I will do what I can," said John.

"OK, put all the weapons and gear in one boat, along with one of the bodies. We put the other bodies into the other boat."

Once that had been done, Ben got out a phone and called his wife.

"Hi, all is well; we are all safe. John and I will be back in the house in a half-hour. Call Michael and tell him I said everything is OK. I love you."

"OK, I love you."

Each of them took one of the boats and headed out. Ben put on his night goggles and shined a flashlight behind the boat for John to follow. They went out about a half-mile and stopped.

"It's over two hundred feet here."

"What are we doing?"

"Getting rid of the evidence. Do you want this dragged through the courts for years? Potentially going to jail?"

"No, but we are innocent."

"Tell that to the prosecutors. They are becoming more and more famous for putting away victims and innocent people. I trust dad will take care of the rest of the issues."

"What about the boats? And won't the bodies float?" asked John.

"No, they won't float. They are taking the rest of the evidence with them."

The captors had left lots of equipment on board. He attached all of it to their bodies, other than the food they'd brought. One by one, he pushed them overboard and watched them sink, making sure nothing rose to the surface. Something might float someday, or someone might find a body at some time. By then, the hint of what happened would be long gone.

"Good riddance," said Ben.

Then he got into the other boat with John, leaned over, and sliced the inflatable he'd just climbed out of.

"The motor should sink it." They watched it deflate and slowly sink.

They headed back to shore. Ben took some rope and wired the engine so it would just go straight ahead. The winds were calming down, so he hoped it would go quite a ways out to sea. He put a small hole in the boat knowing it would deflate within the hour and sink. He just hoped no one had binoculars on it. He aimed it out and let it go. They both waved good bye.

"Thanks for that. Would they have killed me?"

"No."

"Then why take the chance?"

"You mean shooting them?"

"Yes, we both could have died."

"Did any of them look like they had a chance?"

"No," said John.

"Let's just say I knew what I was doing, the risk was very low, and I personally don't think they should live. They will do this same thing over and over again."

"I can't disagree with that," said John.

"We will go in the way I came out. Do you need to change?"

"I'm a little damp."

They got to the house, went in the back, put their clothes in the laundry, went to their rooms via the back stairs, changed into pajamas, and joined the others.

"What happened?" asked Joan, hugging John.

He looked at Ben.

"It's all good. We talked with them, and they left," said John.

"Just left?" asked Joan.

"In their boats and gone."

"Should we call the police?"

"I don't think that will be necessary," said Ben.

"But, don't we want them and who hired them arrested?" asked Sam.

"Do you want to keep asking questions or just trust that it is all taken care of, and they won't be bothering us again?" Everyone stared at Sam. "I didn't think so. I know this will sound crazy, but I am asking you to trust me. No one can ever know what happened here; it's our secret. Because this is connected to national security, the proper channels are taking care of matters. You will not see this on the television. I will deny everything. Michael knows about this and has been involved, as you know. Because of this, I will ask Dad to send security. There will be no show of guns, but until we leave here, nothing will happen. Does everyone agree with those terms and ideas?"

They all nodded.

"Then it's off to bed for me," said Ben.

"You can sleep?" asked John.

"What's the big deal? We were held captive, and we are now free and safe. The kids didn't even know what was happening. Don't tell them," Ben replied.

"Wow, that is ice."

"Just another day at the office."

"Not my office." Everyone laughed.

"Are we safe?" his wife asked.

"Yes. We will have company in the morning when we wake up, to protect us. I know dad will call, so I will wait till that happens before I go to bed," said Ben.

"Ward, have you heard anything?" asked Michael.

"They have Lawson and the phone he was using. What do you want to be done with him?"

"Let him fly from the chopper over the forest where no one goes, or over the ocean. He can join the others," I stated.

Michael agreed. Mr. Lawson would not be seen or heard from again.

Longstreet called Mr. Winters back.

"Mr. Winters, do not call the FBI. I don't think any of us want to drag this out. I am not holding you responsible. If I find out you were part of this on any level, you will pay. Mr. Lawson is being taken care of. When you go into work tomorrow, act just like any other day. Just do what you would normally do."

"Thank you. I am again very, very sorry this happened. I am appalled. Is there anything I can do for you?"

"As a matter of fact, yes. I will text you three dates and times. I would like you to tell your security to let the I.P. address I will send you have open access to your network for two hours on each occasion. I am not taking anything; I don't need to. I just need processing speed. You probably know I could do this anyway. It would just take time and energy. I will ensure your company is treated well by the government."

"Very well, you have my promise."

"I hope we can meet sometime on better terms."

"So do I, have a good day," said Rick.

Michael called the house back and talked with each of his kids before they went to bed.

"Ben, thanks for what you did. I am sorry you had to do that," said Michael.

"I hope whatever you are up to is worth it."

"We will know in a few days. It's almost over."

"Good, I can't take much more of this."

"I beg to differ, but I still feel bad I am responsible."

"No, Dad, you are not responsible, those who did this are," Ben said.

"I assume they paid the price?"

"Let's just say they won't be doing it again."

"I have taken care of the leader, so is hush the word?" I said.

"Yes, and we would like security. Better to be safe than sorry again."

"They are getting on a boat as we speak. Tell the others not to worry when they see people outside in an hour."

"Thanks," replied Ben.

"The least I can do. You know how much I love all of you."

"Yes, Dad, we know."

"You know I would give my life for you and the others."

"Yes."

"I don't expect the same. Endangering your lives was not in the plan."

"What is the plan?"

"Sounds like a question. I ask you that often and I have never received an answer. So, I will use your typical response. Just stuff, Ben, nothing that important."

Ben laughed. "Touche."

"Take care. When this is over, we will all head to the island or up here together. I miss everyone, especially the grandkids."

"That would be good."

"I'm on very good terms with Beth Wilson, your boss's boss's boss. I think I can get you the time off."

"You do move in high circles."

"Too many, I prefer staying under the radar. Hopefully, that will happen in a couple of weeks."

"I have my doubts, but I wish you well."

"Rest well. I will text you when to start watching the news."

"I have been, and I believe I am aware of some of your shenanigans."

"I've had some fun. Night."

"Night."

Everyone was still up at Michael's house in the mountains. He told them all was well, and he was going to get some sleep. They all went to bed. No one at either house, other than Ben and the kids, rested well.

61

Tuesday morning, 9 am in Moscow, 2 pm in Beijing, 3 pm in Pyongyang, North Korea, and 7 pm Monday night in Washington D.C. It's a big planet. Simultaneously, in all those countries but the U.S., the scientists met with the leaders.

In Moscow, the President sat and listened. He had replaced Putin after his reign of terror. The scientist was well respected by the President and the populace. He was on television often and was the President's mouthpiece on scientific issues. The President was told he and the others in the room were now infected. They would die within three days. The virus would be unleashed if they did not agree with the demands. The group of eight talked for two hours, and then four of them complained of chest pain. The President said he'd kill the scientist and his family if he didn't give him the vaccine. The scientist spoke the reality – kill me, and you die.

"Give me the vaccine, and we will take it to the legislators," said the President.

"I am not that stupid. You get the vaccine, and I am dead. Besides, the vaccine does not heal you once you have the virus; it protects you from getting it. I do have an antidote to give all of you. You will sign this document. If you go back on your word, you will die, as will your family. Do you think this is the only virus we have created?"

That shocked the President. He knew he was trapped. Then he grabbed his heart.

"It has started. It will only get worse. I am prepared to die for the sake of the world. Are you willing to die for nothing?"

"You are lying. It's just a poison, get a doctor."

"You can do that, but you will die. It's not a poison. You know me, I don't lie."

The President and the others sat silently, the pain increasing.

Finally, he reached over and signed the document.

"Remember, if you go back on this signing, you will be signing your death warrant and that of hundreds, if not billions, of others. Is your life really worth that? Is your ego really that big?"

"Get the vaccine and the antidote."

The scientist left and returned with both of them.

"You will stay here for twelve hours. Your calls are being monitored. If you say one word about this to anyone for twelve hours, you will die, as will your families."

"You are evil," said the President.

"No, I am not. If you go against what the world wants and needs, you are evil. You will only be harmed if you are disobedient, just as you have killed thousands who didn't agree with you. Don't you dare talk to me about this."

The President shut up.

In China, things went about the same. The President tried to convince Chen, the scientist, that this was not the right path. Chen stuck to her guns. She showed the President videos of what the virus did to animals, along with charts showing infection, incubation, and mortality rates.

"Did you create this?" asked the President.

"That doesn't matter. I have it, and if you don't sign and follow through with what the document says, you will die."

In the end, the President signed.

The scientist entered the dictator's office in Pyongyang, North Korea. Things had not changed in decades in North Korea. People were abused, murdered, and essentially held in slavery. The current dictator was one in a long line of brutal dictators the world had not found a way to eliminate.

"Madam President, thank you for seeing me."

"My pleasure, what is it you want?"

"As we speak, everyone in this room is being infected by a virus that was human-created. It has a 90% infection rate, and those infected die within three days. There is no cure because no one has ever seen this before. I do have a vaccine and a medication that curtails the virus after it has already entered the body, like yours. I do not have it with me, but can get it within an hour."

"Why are you doing this?"

The scientist handed the dictator the document.

"If you want to save your life and those of your family and others, you will sign this. If you don't, you will all die within three days."

The dictator was known to have a temper. She pulled out a gun and shot the scientist dead.

"Go find that vaccine and the antidote," she screamed.

"Where do we look?" asked one of them.

"I don't care, find it." She shot the person that said that.

"Perhaps he was lying."

The others in the room ran from the room. They all knew they were dead. One of them grabbed the scientist's bag. Inside of it was a note.

'I suspect the President will kill me. If that happens, she will die. Upon her death, someone will take over. Unless that person signs the document, people will continue to die. Anyone reading this note will infect anyone with whom they have contact. Below is a number to call when you have a leader ready to sign the document. At that time, they will give you the vaccine. I am sorry it has come to this. Best of luck.'

While a couple of them tried to find the vaccine, the other went to the residence of the next person in line of power.

"I am sorry to disturb you, but I have pressing news. The President will die in three days. She was given a human-created virus, and she shot the person who gave it to her. There is a vaccine, but the only person who knows where it is, is dead. Being in my presence, you now have the virus and will die within three days. Whoever signs this document will live. The President is refusing."

The son of the dictator, twenty-one years old, read over the document.

"Give me the pen. It is time for a change in our country."

"What about the President?"

"For my signature to be valid, she must die. Leave her in the room and lock the door."

"Yes, sir."

"I will call the number and tell them what happened."

"Yes, sir."

"I want our country to grow. This will help speed the process up."

"Yes, sir. Please note, that other than that number, you can't talk with anyone else about this for twelve hours."

"Yes, I saw that, thank you."

"What will you do if your mother changes her mind?"

"She has made her decision. She will need to live and die with it."

"Yes, sir. If I might be so bold, the country will thank you for this."

"I appreciate that comment, and I will need help rebuilding this great nation."

"I stand ready to help."

The day they'd been building up to finally arrived. Mary said she would like to join Clint, Michael, and me for the D.C. meeting. We all snuck into the city the night before and were staying at four different hotels. The session was closed and was at 9 am. I was not invited but would be waiting outside the hearing room.

The scientists in Russian and China sent me copies of the signed document shortly after they'd been signed, and the Presidents had received the vaccine.

I was nervous when I didn't hear from North Korea. I had mixed emotions when the word came through the document was signed by the new President, the elder son of the current one, who shot the scientist and was already dying. She'd had health conditions prior to the virus. Clint was saddened by the loss of a colleague but knew the scientist who'd been shot would say it was worth it.

The following morning in the United States at 9:00 am, Clint, Michael, and Mary entered the closed session room. They knew the other countries had all signed. No devices of any kind were allowed in. The walls were made of metal that could not be penetrated. There was one phone that was a landline. There was also one computer that projected on a screen, so everyone saw what the user was doing.

"Good morning, we are glad you are here. Once again, you called this meeting, so we will turn it over to you. We hope it is more good news," said Senator Dickinson.

This was a joint committee, so there were eighteen people in the room. They all showed up because of the fame Michael, Clint, and Mary achieved from their recent activities with malaria.

As Michael started talking, Clint got up and put a can with the virus over one of the ventilation vents. It was blowing air. He knew they'd be infected within thirty minutes.

"Chairman Dickinson, we appreciate your willingness to meet with us. We happen to believe what we bring is very good news. You may differ. May I use the computer?"

"By all means."

Michael was essentially stalling for time to let the infection take hold. He spoke about the advancement of AI and GE, Genetic Engineering. A few of the people asked questions to gain understanding. Michael then talked about their concerns on both levels and what could potentially occur if AI or GE got into the wrong hands. He showed graphs of infection and death rates globally. Then he pointed out what advanced AI could do. As if on cue, the air being sucked out stopped, and the blowers of air coming in, increased. All of their ears popped.

"As you know, I have created a new level of AI. One thousand six hundred attempts have been made on my computer by the greatest minds on the planet. No one has come close. As you have no doubt heard, some of those computers are now useless, including some of our governments. You undoubtedly have heard about electrical outings at sporting events, economic issues with banks and the stock markets, and other unexplained events. I did them all. This is a minor inconvenience I created to show what advanced AI can do. I can't allow that to happen. I am unwilling to give that power to any government for any amount of money. AI in the wrong hands would lead to destruction beyond your imagination."

"Dr. Longstreet, you are aware we can have you arrested for what you have done, and simply take your software."

"Yes, Senator, I know what you can do. I will let Dr. Williams speak first. I will end by saying if you do as you suggest, that will trigger my computer to set off a series of commands that cripple the banks, make the military's radar useless, and shut down permanently the three major computers the government uses. A pre-recorded message will be broadcast on all networks, stating what you did and what you caused. Your careers will end, and probably your lives. I don't think America will be pleased. If I do not tell my computer that all is safe in the next 30 minutes, the same will happen."

"How dare you."

"Don't insult my intelligence, and I won't insult yours. Let's hear what Dr. Williams has to say, and then we will let you know what we want," Michael stated with some anger in his voice.

"I will be brief," said Clint. "While Michael was speaking, a virus Mary and I created was released in the room. Don't bother to cover your mouths. It's already in you. Here are the graphs."

Clint put graphs on the screen that showed infection rates of over 90%, incubation of fewer than four hours, and death at seventy-two hours. All the Senators went white. A few got up to leave.

"Sit down. First, Michael has locked the doors. AI at work once again. Secondly, if you walk out of that door without the vaccine, you will infect virtually everyone you contact. Just so you know, we do not have the vaccine on us. We aren't that stupid. Thirdly, you will all die within three days. It will attack your heart, and you will die of a massive slow heart attack. We do have a vaccine, and a medication to stop the virus once it's in a body. We will give you both."

"You are mad," said one of the Congresspeople.

"We choose to believe we are the only sane people in the room. All of you are complicit in allowing countries around the world to develop GE and AI without boundaries. We have come to set those boundaries. As you will see, we ask for nothing, zero. This is about the world, not us."

"Before going further," Mary chimed in, "you need to know this exact same thing has taken place in Russia, China, and North Korea. Each of their Presidents has signed what we will ask you and our President to sign. In North Korea, the President shot the scientist making the presentation and is probably hours away from death. Her son has taken over and signed. He also made it clear that he will be doing things differently in North Korea. He announced to us that he is taking down all nuclear forces and inviting our investigators to watch. This is something you have been working on for over sixty years. We accomplished it in a day. We do not care if you take credit for everything. We are sure you know Tom Armstrong. He will be writing about what has been taking place, including the kidnapping of Tom's agent and Michael's son, who works for the CIA, by his own government. So, be careful what you say – that is, if you are alive. The clock is ticking."

"What do you want?"

"Glad you asked Senator," said Clint. "We want the world to be safe from AI and GE manipulation by governments or companies. No one has the right to that much power."

"We would never allow that to happen," said one of the Senators."

"Of course not, like you didn't allow Japanese Internment camps, segregation, the use of prisoners and soldiers to test military medications and drugs, the manipulation of elections, the use of information for insider trading, the ordering of me by one of your generals to create this virus, the bombing of innocent civilians in war-torn countries, the stupidity of not slowing down COVID when you knew exactly how to do it, or the invasion of people's privacy for no reason. I have dozens more if you need them. Before you open your mouths, let me warn you, we have items we can prove that would bring virtually all of you down. If you challenge us, your careers are over. If you continue to behave as many of you have, your careers will be over, and you will be in jail," said Michael.

"This may sound like we want to take over, a coup. We have no interest in power or money. We want a safe world where people can prosper and not live in fear. If that is not what you want, please say so now," said Michael. Silence.

"We are handing out the demands." Just so you know, the President has been infected. He has three days starting six hours ago. I suggest you call him; he should be feeling it by now. If he checks his secure email, he will have the same document you have and that the leaders of Russia, China, and North Korea have signed. If the President would like to join us, he can. We will put him on the screen," said Clint.

The Senator made the call. "Yes, put him on the screen." Michael did.

"Mr. President, I will assume you have some pains," said Clint.

"Yes, they have sent for the doctor, it's in my chest."

"We have infected you with a virus. It will destroy your heart in three days. There is no cure. Anyone you breathe on will die within three days. You also have in front of you the document we need you to sign. The Presidents of Russia, China, and North Korea, have already done so. If you wish to call them to validate that, you can," said Mary.

"We will make this very simple. The documents are written in humanease, not legal or politicalease. There is no negotiation. Every country that does not sign on will suffer the consequences, starting with the leadership. You cannot hide. If you want to test us, go ahead. What you have seen

over the past few weeks are small potatoes. I will give you
the cure and the vaccine when you sign. You will have until
you die, that is about two days. If you need to call other
legislators to get this done, do so now," Clint said, his
adrenaline soaring through the roof.

"As you will see, I will turn the software you all want
over," said Michael.

Global Treaty on Biological and Chemical Warfare,
Genetic Engineering and Artificial Intelligence.

1. ALL genetic engineering will have a minimum of four
countries represented in all the work. This means all
Level 4 laboratories will have a minimum of four
nations represented. Those representatives will be
vetted by Patricia and three other networked
computers.

2. Four computers will be built under the direction of
Michael Longstreet. One will be in the United States,
one in Russia, one in China, and one in Australia.
They will be networked together and will oversee
the work of Artificial Intelligence and Genetic
Engineering. There will be a committee of seven
that will oversee the future programming of these
computers. This work will be done through the
auspices of the United Nations. All countries signing
this document will donate .25% of the GDP to the
fund to build, run, and oversee these computers.
This is the equivalent of about 250 billion dollars.
They will have access to all information on the
planet.

3. Elections will be held in every country within two
years. Those elections will be overseen by the
computers to counter fraud. All debates will be fact-

checked. If a candidate lies, their microphones will be turned off, and Patricia will let the audience know the evidence behind the lie. None of the computers are politically oriented; they are truth and life-oriented. If anyone in an elected position has been part of corruption, we suggest you resign.

4. All countries with a GDP of over 1 trillion dollars the year before this document is signed will donate .5% of that year's GDP to a fund overseen by the United Nations that will be used to bring fresh water to everyone on the planet. A rough estimate is 450 billion dollars. If that does not cover it, another .5% will be collected.

5. Corruption on any of the above, on any level, will lead to twenty years in prison. The computers will provide the evidence.

"That is it, very clear and very simple. We do have one more document we require all nations to sign. Most have nothing to do with this document, but we want assurances. There is a list of names on this legal document, written by lawyers, but also in plain English. By the way, all documents have been translated into the language of each country," stated Mary.

"The leaders of your countries will be signing this, pardoning all those listed from any crimes committed up to this point and with any crime in the future as we protect ourselves, our families, and those associated with the work we have done and will do relating to this document. I will be blunt. An attempt to harm any of us will be met by swift and severe punishment. At the hint of an attempt, the computers will act," Michael added.

"Four days from now, we will be going on international television to let the world know what has happened. You have a choice. You can lead the world and present this, or you can let us tell the world why you died. We want no money or credit for this. If you wish to claim all the credit, please do. I think you know the truth will come out. Tom Armstrong will be releasing a book telling the world how all this unfolded. You can be on the right or wrong side of history," Clint said.

"In case you are wondering, the computers do not have the ability to launch major weapons, but they will be given the power to blow them up if their targets are unrighteous. The computers will not stop wars. They will prevent the death of innocents and the annihilation of countries. There are only two reasons to launch a weapon. The first is offensive in nature, to destroy an opponent. The second is defensive, to protect your country from those offensive weapons. Weapons will not be allowed to be launched for offensive purposes, so, in the end, there is no need for missiles or bombs. Those who seek that level of destruction will reap what they sow. Our hope is that it will not take many examples for people to get the idea," Michael added.

"Unless you have questions, we will depart. We will not tolerate any interference. Please understand, if I or anyone on the list is harmed or held hostage, you will die, as will your families, and probably millions of others. I see that a few of you are already feeling the effect of the virus. Without the cure, you will be dead in three days," Clint stated firmly. "Are there any questions?"

"Patricia, unlock the doors."

"Who is Patricia?" asked Senator Dickinson.

"The leading AI of the future," said Michael. "By the way, I will allow whatever people are on the to-be-created committee we mentioned to look at my software to let you know I have done nothing but seek to protect the world. If you sign the document, call this number or email this address, and we will heal and vaccinate everyone infected. We suggest you not leave this room or the Oval Office until you have made a decision. If you choose not to sign, we will know, and history will unfold or implode. The world is literally in your hands, as are your lives."

They left the room. The President called the other leaders of Russia, China, and North Korea who validated what had been said. He learned from the new leader of North Korea what he hoped would happen. That brought a smile to the President's face, and he was already gloating internally that the North Korea problem would be resolved under his leadership. The leaders all agreed to make public announcements to the world together. They would meet in Germany in two days. They wanted to preempt the scientists being on TV.

I joined them in the hallway. Clint gave me a thumb's up. The Senators and President called and emailed Clint and said the documents had been signed.

Mary went back to the Senate room to vaccinate everyone and get the official document. She also gave them the "antidote" for the infection. Clint and Michael went to the White House and vaccinated the President and those around him that had been infected and gave them the antidote.

The helicopter landed on the White House lawn. Mary met the rest of us there.

"Do not have us followed or tracked. If you shoot us down, the doomsday scenario is triggered, and your world will disappear. Am I clear?" Clint asked the President.

"Very." He got on the phone and gave the order.

"I suggest if and when you meet with the other leaders, you sign a nice pretty version of this document for the world to see. I know there will be a press conference at that meeting."

"How can you know about that meeting? I just talked with them?" said the President.

"Would you like a transcript of that conversation?" Clint asked.

"I guess I should know better. No, that won't be necessary."

"One more thing and the other leaders have been told this. At the press conference, there will be no lying. If you or any of them lie, televisions around the world will not only state the fact but will also show the evidence to back it up. Do you understand? We are tired of lies."

"Yes, I understand."

"Good. I wish you well. One more thing. Please have the secret service or police clear a landing spot for our helicopter just south of the reservoir by the baseball diamond field in Central Park, NYC. Clear all air space; I don't want issues. Remember, if anything happens, the destruction is on you, and our computer will tell the world."

Michael walked into the room. He'd been talking to tech people at the White House about next steps.

"Hello, Mr. President."

"Hello, Dr. Longstreet. I hope this works."

"It will." He turned to leave.

I called Linda and told her to be at the northern baseball diamond south of the reservoir in thirty minutes.

"Why?"

"Don't ask questions this time, just be there," I said, feeling bold.

"I'm in a meeting."

"Leave the meeting, be there."

"OK."

We took off, flew around the city, hopeful the President had added to the future of the country. We landed on the baseball diamond. About ten police and men in black suits making a large circle as we landed. I opened the door and motioned Linda to come. She ran to me.

"I'm going to the apartment. I will join you at the house in a few hours," said Michael.

He jumped out, and we took off. An hour later, we were back at the house in the mountains, celebrating, sharing with everyone what had just happened.

The TV was on, the kids watching a movie. We changed it to the Breaking News, which was on every station.

"Good afternoon America, I am Lester Holt. We have just learned the governments of the United States, Russia, China, and North Korea, have signed a document that will change the course of history. All we know at this time is that it involves Artificial Intelligence and Genetic Engineering. We have also learned that the President of North Korea is dead, and her son is the new President. He is stating he wants to end the animosity and is willing to denuclearize his country. He also stated the wall between North and South Korea would be gone within the month. We have no idea how this happened, but we will keep you informed. There will be a press conference in Geneva in less than forty-eight hours by all the leaders, and those of other nations who are going to sign the document. It will be televised worldwide. If what they say is true, this will be one of the most important documents signed in world history. Set your DVR if you can't stay up. This is Lester Holt and the whole team reporting. Take care of yourself and of each other."

We waited for Michael to arrive before clinking the champagne glasses. He showed up early evening, and we toasted a job well done. Everyone toasted Patricia, Hypatia, and Simon for their work. Patricia said they were all grateful they could help.

I pulled Michael, Mary, and Clint aside when things were settling down.

"I'm just curious. You know I, and others, have a concern about AI computers taking over and doing bad things. What if that does happen? What will you do?" I inquired.

They all looked at each other and said together, "Call you." I smiled, nervously, groaned, and walked back into the house. They all laughed. Deep inside, my gut didn't feel good.

The international news conference was held with 186 countries in attendance to sign the official document. The countries that didn't sign were left to fend for themselves. They all ended up signing within the year, and the leaders were sent packing.

Epilogue

This section explains what has taken place since the event some three years ago. The first two editions were printed without this part.

During the first few months, there were small battles. Threats were made by various countries. Patricia took care of those quickly. Six dictators were relieved of duty.

Over the next two years, truly democratic elections were held in every country. Some chose to remain communistic with authoritarian rule. Most just wanted a non-corrupt government and a level playing field. Patricia, Luke, Ollie, and Tana were the four AI computers that ensured corruption was very small.

Nuclear weapons were slowly disassembled; there was no longer a need for them. The advances in medicine were astounding as scientists from around the world worked together and had access to the AI computers.

In America, 126 of the 435 representatives either resigned, retired, or did not run for reelection. Corruption had run deep. Twenty-eight senators did the same.

As the computers increased their knowledge, advances in agriculture and renewable energy developed quickly. The panel overseeing them worked well together.

Clint retired to his farm with Luly and continued to enjoy using his mind and playing with the grandkids.

Michael helped set up all the computers and was on the panel for the first two years, then he left that work, married Sonya, and disappeared to a foreign mountain. I knew how to find them.

Mary took a teaching job at Stanford. Clint made sure she was well taken care of financially. She had obtained numerous patents and had another child.

Ward and the others oversee the security of the computers. There is a five-mile radius around each of the buildings where the computers reside. Anything entering does so at its own risk, including things that fly. There is a one hundred mile no-fly zone around the facilities. Surface to air missiles were installed before the computers were built. They have only been used twice – with complete success.

Linda and I continue our journey. Her kidnapping changed us. We are now spending much more time together and not seeing anyone else. So far, so good.

With AI overseeing most things, I am running out of conspiracies to write about. They still exist, and I am on the lookout for them if you have ideas. I had been working on election issues before all of this. With Patricia and the others, those are history, as is the book.

There is still crime, poverty, disease, small military conflicts, sports, concerts, school, and the rest. But I do sleep better at night and am thankful the world is actually working together on many issues.

List of Main Characters

Tom Armstrong- Investigative Reporter
Dr. Clint Williams – Geneticist –
    Luly is his wife
Michael Longstreet – AI expert
    Ben, Joan, and John are his children
Mary Soderstrum – Clint's colleague Ward – Head of security for Tom, Clint, Michael, Mary, and their families
Linda – Tom's literary agent
Beth Wilson – Director, CIA
General Fleming – Head of Biological and Chemical Weapons for the military
John Spencer – Director of the NSA

www.ingramcontent.com/pod-product-compliance
Lightning Source LLC
Chambersburg PA
CBHW071532200326
41519CB00021BB/6460